1.⁵⁰

A FIRESIDE BOOK
PUBLISHED BY
SIMON & SCHUSTER

New York   London
Toronto   Sydney
Tokyo   Singapore

TIME
IS AN ILLUSION

CHRIS
GRISCOM

WITH WULFING VON ROHR

ACKNOWLEDGMENTS

For this revised American edition, a draft translation by Janis Kaser of the original German edition was used. Her support is much appreciated by the authors.

Fireside

Rockefeller Center
1230 Avenue of the Americas
New York, New York 10020

Published by the Simon & Schuster Trade Division

FIRESIDE and colophon are registered trademarks
of Simon & Schuster Inc.

Designed by Barbara Marks
Manufactured in the United States of America

10 9 8

Library of Congress Cataloging-in-Publication Data
Griscom, Chris
    [Zeit ist eine Illusion. English]
    Time is an illusion/Chris Griscom, with Wulfing von Rohr.
    Translation of: Zeit ist eine Illusion.
    "A Fireside book."
    1. Light Institute (Galisteo, N.M.)   2. Time—Miscellanea.
3. Griscom, Chris.   I. Rohr, Wulfing von.   II. Title.
BP605.L53G7613   1988
299'.93—dc19   88-15806
ISBN 0-671-66334-8

Originally published in West Germany by Goldmann Verlag as Zeit
ist eine Illusion, 1986.

This book is dedicated to
Wulfing von Rohr
"For the beginning . . ."

CONTENTS

Foreword: How this Book Originated   7
Introduction   15
1. Born Open   17
2. Summer Solstice in South America   35
3. Unveiling the Higher Self   51
4. Echoes from Other Dimensions   75
5. Time Is an Illusion   101
6. Our Choice: Victim or Creator   113
7. Manifestation in Africa   125
8. The Female Explosion   141
9. Honing the Energy   159
10. New Mexico—the Magic of Light   173
11. Death Is an Illusion, Too   185
12. A Glimpse into the Future   209

FOREWORD:

How this Book Originated

Dancing in the Light is the book of superstar Shirley MacLaine. Through this book—at its publication a best seller—many spiritually interested people were made aware of a remarkable woman with unusual capabilities: Chris Griscom.

It was the same with me. After reading Shirley MacLaine's first spiritually oriented book, *Out on a Limb,* I got a copy of her newest. In this book, *Dancing in the Light,* she mentioned that Chris Griscom, with the help of an unusual style of acupuncture, facilitated glimpses into other time-and-space dimensions, and past lives as well, in the process enabling conscious contact with one's "Higher Self." At this point I was on my way to New Mexico, the "Land of Enchantment," as it reads on all the license plates.

I had the chance to visit for two weeks with Chris, her co-workers at the Light Institute, and her family, in Galisteo, a small town south of Sante Fe. For two weeks in February 1986, I was a "participating observer" during the sessions, in private meetings, during their routine day-to-day life, and during special celebrations.

One begins the sessions with an "emotional body balancing." A spiritual "opening" is accomplished with a combination of precise, gentle touches at specific points. The solar plexus is stimulated to release the astral body and open the auric field.

After this opening the client is guided, while still on the massage table, through different childhood states. This contributes to an initial clarification and intentional self-scrutiny, bringing about a state of mind essential for the following sessions, which concentrate on pictures and experiences from other times and spaces.

I reexperienced myself as a four- or five-year-old boy, exhilarated, carefree and joyous; thereafter as a ten- to twelve-year-old youth, who had already become more serious and exerting; finally as a sixteen- to eighteen-year-old young adult, who was controlled, disciplined and situated right on the threshold of life. As the session progressed, this reexperiencing allowed me to see how the various vibrations of each time period were relived in almost all body cells. This in itself was very special.

After this preparation, there normally follow three past-life sessions, each around three hours in duration. In this respect I was pleased to notice that Chris and her colleagues do not insist on calling the sessions "guidance into past lives." Every client prefers to label his or her experiences in his or her own way. It is important for the client to determine for himself how the recalls relate to him as they emerge from the self in the form of pictures, feelings and thoughts, and the client can see their relatedness to present life circumstance.

The past-life sessions are begun with the same techniques used in the preparatory session. Although one's consciousness of the body gradually becomes dimmer, a light-filled body awareness remains, with Chris's facilitation. The client will be diligently guided "inward" according to occurrences during the sessions.

Throughout the duration of the "journeys," I clearly felt energy currents in my body. This energy was cleansing and invigorating. I repeatedly had the sensation that my astral body, at a distance of about four to sixteen inches from my body, was worked through with a feather duster, or "brushed off," so to speak. During the sessions the presence of other beings besides Chris were not only felt but in two cases visible: I saw ordinary faces looking at me.

The initial sessions were conducted by Chris's daughter Karin, who is one of the colleagues at the Light Institute. The experiences were noted down so that a permanent record could be kept. A few session extracts should suffice:

▶ I experienced myself as a horseman on a war horse, who delighted in visualizing his victories. I saw myself in a spiritually dull state of mind in a tavern circa 1700; afterward as a gooseherder who had found tranquillity and joy through harmony with nature. I recalled myself as a young Indian boy of the wide plains, who spoke with a buffalo—a very intense experience of harmony between nature, animals and man. This Indian boy found himself again along the edge of a forest as an old medicine man, who had become completely distraught over his visions of the vanishing of buffalo herds and the natural way of life for the Indians, and finally the decline of the last harmony between nature and man.

▶ The next scenes showed a young woman caring for the wounded on the battlefields of the American Civil War. At a later time she lay as an old woman on her deathbed. To the astonishment of her assembled family she suddenly revived, radiant. She saw a light coming in at a slant from above, guiding her upward and within. This was death as a light-filled experience, without horror. "Up there" the old woman encountered the Indian medicine man and the buffalo. Both persons fused into each other. At this same time I was struck with an insight: it seemed as if the medicine man purposely returned as the helpful woman to play a part in relieving the suffering he had already foreseen as a medicine man.

▶ Another death experience: I saw and felt myself lying "dead" on the ground in a basin. One after another appeared a buffalo, a bear, eagle, coyote. The buffalo nudged me with his nose and said, "Wake up, little brother." At that point I "awoke" and as an iridescent form, raised myself.

Whether or not these incidents were past lives isn't really pertinent. Much more significant is the direct, firsthand experiencing that penetrated through every cell in my body and that even now remains recallable, relivable.

Perhaps also of interest is that I later experienced a significant corroboration of some pictures from the sessions. At the Indian Market at Santa Fe I found a small ornamental heart surrounded with a ring of feathers, exactly as I had already seen with my mind's eye in the session. On a visit to a bookstore, an animal collection revealed itself as the totem animals of Indian astrology. At home in Munich lay a two-year-old psychographic drawing from Carol Polge of England, depicting the exact face of the woman who had cared for the wounded.

After a detailed initial interview to ascertain and identify emotional and spiritual themes, Chris began with the acupuncture. She stuck fine gold acupuncture needles into strategic spiritual points. These points, called "windows to the sky," are not the usual acupuncture points.

Chris explained that as the sessions proceeded, completely new levels of energy would be opened up. This allows a conscious experiencing of other dimensions, times and spaces through one's emotions, reason, brain and higher self.

The sources of problems become clear. There is a catalyzing effect between insights and solutions. Along with the surfacing of events out of past lives appear glimpses of their connections to future events or messages, similar to the results of "channeling" phenomena.

I noticed some external phenomena at this state of our work. I felt, long after Chris had ended her session, as though the acupuncture needles were still intact. She explained that a spiritual helper, a Chinese doctor, was helping her and had set "astral needles" in the astral body. During these sessions the energy currents appeared again in my body, manifesting as a sudden vibration of my limbs and leaving, as before, cleansing and invigorating effects.

Although my Higher Self repeatedly showed itself as an iridescent form in the preparatory session, I perceived it now in many other aspects: as an eagle, snake, a red stone column with a yellow corona, as a female being clothed in blue, as a royal couple. Very frequently the melting together of two aspects occurred, which were like "yin" and "yang," or masculine-feminine opponents. The most moving of these recalls was the scenery around a pyramid under construction. At that time the pyramids served as energy structures, which took in light energy from above and passed it along. I saw how a Pharaoh stepped into a light-filled space ship on the tip of the pyramid, melded with it and disappeared.

I saw light spirals become activated in myself so that I

could leave the body. I saw how I had been admitted into light-cities of the hereafter, and much more.

These experiences have little importance to me as something titillating or sensational. On the contrary, as the sessions progressed, so came the deep conviction that much creative strength slumbers in everyone. I realized that every individual has the chance to effectively search for the purpose of life and to realize certainty in his life.

My two weeks spent at the Light Institute in Galisteo with Chris and Karin Griscom resulted in a new outlook on life and a more conscious connection to an immense source of inner strength on many levels and in different dimensions, simultaneously. I realized that when one has power or is associated with power, one's motivation is decisive. One doesn't want to become pulled into power abuse or into a dangerous application of power, be it even very subtly.

Finally one must question: *Who* experiences? Regardless of how fantastic the experiences may be, this fundamental question remains. The sage Ramana Maharshi said, "The true self is continuous and unchangeable. The reincarnating ego belongs to a lower level, mainly the world of thought. It will be transcended through self-realization."

In this light I see Chris's work as a valuable aid to developing one's consciousness without taking it as an "absolute" path to a purely spiritual consciousness, as some teachings of spirituality profess to be. My personal impression of this woman was, Wonderful, this is no guru type. She does perceive herself as existing within a continuous line of former lives reaching back to Atlantis, she does feel carried by a mission to reveal to people the spiritual powers inherent within them, but she still manages to keep her feet on the ground.

Chris was born in 1942 and has six children between the

ages of twenty-three and four years old. Bapu, the youngest offspring of the family, was brought into the world in the sea by Chris alone. She also works in her greenhouse, accompanies her son Britt to the ballpark and cheers him on, and quickly helps with acupressure when he or another hurts himself. Despite achieving reknown as Shirley MacLaine's acupuncturist and despite her astounding abilities to guide people into other dimensions, she appears to have remained a perfectly normal person.

After my time spent with Chris, I was determined to bring her even more into the public light. Out of a TV special for the German national TV chain ZDF, the idea was born to publish a book. It is Chris's very first book and results from many tape-recorded meetings and interviews done during my trip for this purpose in June 1986.

We naturally had to restrict ourselves to a selection of important topics from her wide-ranging life. It is a selection that nonetheless makes clear why Chris is considered one of the most significant "New Age" and "medicine" women of today.

Her newest project is the Nizhoni School for Global Consciousness, which focuses on the spiritual education of young people.

Wulfing von Rohr

Time is a dance, a flickering gesture of the mind,
a curving, arching orchestration of the soul.
It bears the three faces of subjective truth
yet it can be molded, squeezed or stretched
by cause and effect
until it thins itself out to become
the impenetrable membrane
which binds the third dimension.

Place the consciousness on the past and it comes
alive, project to the mist of the future and it suddenly
takes form, breathe into the present and it washes over
you and fills you with the sweetest nectar of life.

We, ourselves, stand mesmerized beneath time's
looming shadow, afraid to remember space, and thus
our fate lies limp along its horizontal wall—not daring
to dream the great beyond!

INTRODUCTION

The structure of our lives is so indelibly inked with the constricture of TIME that we have virtually sealed off the perceptual threshold to our multidimensionality. The past haunts us and drives us into a threatening future—which rushes in with such lightning speed that we feel unprepared and totally incapable of coping. The present is a shadow zone so cluttered with past and future energies that we never actively engage ourselves, except when acute pain or pleasure snaps the precarious thread of our attention and we abruptly bump into the now.

Tragically, the now for most of us is too filled with meaningless noise to allow the senses an integral experience of peace. The noise blocks out the vibrant sound of our true self. When we are disconnected from that divine self, our experience loses meaning and we are set adrift in a sea of confusion and repetition. Cognition of purpose is crucial to our evolutionary process. Indirectly, the entire scope of reality changes when we can view experience from the vantage point of our true purpose. In the light of this all-knowing intentionality of our soul, negative experiences take on cohesive meaning as tools that we are utilizing to approach the point of clarity and growth. Our recognition of this process helps to wean us from the addictive emotional body, which feeds itself upon daily dramas, constantly seeking to repeat the feelings by which it identifies itself.

On the flat surface of daily life, there are clues to energy levels of awareness that might free us from our unconsciousness treadmills. The light shines too dimly into the cracks of illusion to allow us the bliss of wholeness. The routine of our lives acts as a buffer against the possibility of exploration outside time and space. Yet, there is an inkling of an answer as to who we are, which awaits our initiating a journey within.

When we focus our consciousness on the inner reaches of our being, the veils of separation caught in the tapestry of time dissolve. We discover that our bodies are spin points, virtual thresholds between many dimensions and realities, which speak directly to what we are creating for ourselves, today.

Time Is an Illusion is a whisper about how I found the light between the cracks and came crashing into our multidimensionality.

I truly hope it will inspire you to explore yours. Already, from so many German readers, I hear a resounding "Me too!" (This book was first published in Germany.)

I feel certain you will find yourselves here too, and that this book will bring you a cosmic giggle of recognition and an impetus to go beyond.

Chris

BORN OPEN

1

Each one of us is born clothed in full spiritual regalia. Wrapped within the essence of our life's purpose, we hold the meaning, the choice, the ending to the story of any particular lifetime. We need never hunger to become more spiritual. Spiritual energy is not apart from us, it is not in front of us, it is not something to be bought or gained—it is already our memory, it is already our beginning. Though we step into the vacuum, into the illusion of time on a continuum, and forget to breathe in, or look back, or experience around us our innate spirituality, we can never truly be separated from it.

But those timeless, multidimensional aspects of our being are driven rapidly from our consciousness. Even from the first experience of birth, we are confronted with the bombardment of sound and color and cognition of pattern to all of the vibrant senses that distract us into our experience in this body, this lifetime. The tendency to close the door to that multidimensionality, to turn away from the whisperings of the soul, increases as our attention becomes fixated in our little bodies and in this life.

It is true that the brain can only record the strongest of signals one at a time, and that, perhaps, is the basis of limited consciousness wherein we begin to perceive time as a continuum, one thing then the next then the next, rather than perceiving through the hologram of a brain that has within its depth the capacity to imprint the flash of all contingent realities that emerge in a medium of simultaneity. The experiences we have from the moment of our first explosion into being that comes from conception—throughout the maturation in utero, into the birth canal, and from the fresh new vehicle of the body—influence the way our brain perceives. If we look at the emotions and experiences, waves of thought and feelings that come into contact with that first procreative medium from the father and the mother, we can understand better the point of reference that each of us has in terms of perceptual reality.

I, for instance, entered the body, the womb, of a woman who had been told that giving birth a second time would be a death experience for her. With a certainty reverberating in a body in as poor condition as my mother's, from the beginning she set the intention of her will to outdistance the possibility of death, which was so omnipresent. She did this by focusing her attention on pure pranic energy, which is the essence of the life force itself. I actually experienced this energy in the womb and remember the sensation of its fluttering particles of light.

At the same time my consciousness remembers this light, this energy, when in her seventh month of pregnancy my mother succumbed to toxemia and kidney malfunction. The poisoning that affected her was also available to my brain patternings and, fascinatingly, years later when I had spontaneous uncontrollable out-of-body experiences at the age of nine, the patternings that precipitated those out-of-body experiences were red and blue lights in concentric circles and lines, actually the same patternings I experienced in my in utero state. So the disruption of brain pulsation may participate in causing spontaneous out-of-body states, especially since very often out-of-body experiences come during operations or accidents when the brain pulsations are disrupted. It's a very interesting correlation between the two.

I remembered those experiences in utero, although my mother later told me of the conversations she, my father and the doctor had had about who was to be saved in case only my mother or I could survive. I, in my precognitive, nonlinear consciousness within her body, perceived those conversations and, in fact, during the moments of my birth I experienced profound guilt and anxiety and pain that she was to die, that I might live. Even though those emotional imprints were without thought as we know it, the prophecy of the doctor was indeed played out and I was delivered prematurely while my mother, during my birth, experienced heart failure, passed into death, and was only later revived. She was therefore unable to connect with me for the first six weeks of my life. We each lay solemnly alone in our beds, weighing the choice to live or die. It was her pure will that gave me life, and even in the passage of birth, I experienced and recorded the struggle of death. I weighed three pounds at birth, and I remember how my father loved to tell me that he could hold me, shriveled as a prune, in one hand. The experience of dying and birthing simultaneously, which my mother and I shared, perhaps allowed me the foreknowing of those two patternings that occasion every birth and every death, and literally go on perpetually on a cellular level within our physical vehicles. At that point there was certainly the imprint of confusion as to which way to go—into life or into death. Because I experienced them both, perhaps I was "born open," more profoundly open—flooded with all the future living brings—that ultimately circles back to death, and therefore holding a frame of reference that has vacillated my consciousness inside and outside the body.

The events of my birth created some internal familiarity with passage to death and passage to life and I, indeed, experienced several close encounters with death during my first five years. I was seriously ill with whooping cough, diphtheria and scarlet fever. During each of these experiences, my consciousness reenacted those initial passageways back and forth and in and out of body. These experiences enabled me to cross the bridge into other dimensions and to vacillate between this world and worlds beyond. Occasionally I found myself outside the body on astral levels as well as in more ethereal levels, which lead closer to the center of the soul.

I remember very specifically my experience with whooping cough because of the sound and thus the vibration of the whooping cough ricocheting through my body, producing an interesting sensation. And I remember lying on a table in a white room while my parents spoke to a man in a white coat behind a glass door. Although I couldn't directly hear their conversation, I perceived the doctor saying to them, "It's possible your child will die." I felt the constriction around my mother's heart and a tremendous disruption in my father's auric field.

At that point I felt a forceful pull in myself and floated first out of my body, back in, and then out again. Something in me flew upward and wanted to leave the body for good. The experience of moving in and out of my body had seized my complete attention. It felt like being a large white balloon swelling with air, floating upward, then deflating and sinking. The experience was very liberating, though the indecisive leaving and returning was very strenuous. There was some hesitation on staying with this little body.

There were many magical moments during the first five years of my life. These experiences with the other forms of life, other beings, and some moments when rocks spoke to me all seemed to be a normal part of life. I was deeply entrenched in the wonderment of nature, yet I found the interactions of people the most fascinating of all. I loved to sit and just listen to people speaking to each other. Although I was extraordinarily verbal at an early age, my mother always pointed out that I used language in a strange way, choosing my words for their sound rather than for their literal meaning. With time, I observed that my family and their friends began to consider me a bit odd. I supposed that my odd perceptions disturbed the grownups, along with my habit of expressing myself very bluntly; I was too irritatingly direct.

Around the age of five, a psychologist friend of our family took me to UCLA to be tested. We entered a large, austere, intimidating building, up many stairs and down long corridors, into a special testing room. On one side of the room was a special one-way mirror, whereby the psychologist and his colleagues could observe the room without themselves being seen. Because of my sensitivity, I knew there was someone on the other side of this mirror. Instead of proceeding uninterrupted from one task onto another, I would repeatedly turn suddenly around, in an attempt to "catch" any of the people on the other side. I didn't know whether the people on the other side were my friends out of elfland or other beings, or just real flesh and blood people. I only knew that behind this mirror were living beings, clearly perceptible although visually obscure. I remember approaching the mirror and my reflection, trying to push my body through and connect somehow with the energy currents on the other side. The game became delightfully intriguing, and I soon forgot all about the "peg-in-the-hole" exercises I had been given.

Suddenly a woman entered the room, appearing displeased that I had not finished my test. I compliantly sat down and quickly completed it, but from this point on, the feeling that I had failed to satisfy the grownups tormented me. They appeared very serious and distant, leaving me with simply feeling that they had concluded something was wrong with me. At that time I did not understand why I was being taken there. I only sensed that it was very important for me to somehow please my father's friend, and I complied as well as possible with his instructions. But I did so within a holographic latticework of incoming perceptual data that simply exceeded the adult linear model. The exercises were IQ tests, which were not capable of measuring brain function on deep levels. Until we learn to map these multidimensional mind potentials, we will continue to interface with our children from limited and tragically shallow octaves of communication.

When I turned nine we moved to the countryside on the outskirts of Los Angeles. There was a walnut grove nearby that became a very important spot for me. Between the trees swept a sea of wild mustard, with very bright yellow flowers. The mustard grew to over three feet high, so dense that it appeared to be a yellow blanket. For several years I remained shorter than the mustard blanket and could stand surrounded by the fields without being visible.

A large part of my childhood was spent within these mustard fields, creating my own world, crawling around, eating their roots, digging holes and hiding myself in them. Yellow stimulates the nervous system, and my sensory perception was wonderfully enhanced in this magical world. I experienced all manner of life forms directly and personally about this time. However, a dark cloud came over my idyllic world.

During a tonsillectomy for which anesthetizing gas was administered, I was catapulted out of my body and had the shocking experience of viewing it from just a few feet above. Always before I had simply left the body behind, identifying with the sensations themselves rather than any awareness of my body as a whole. This new perspective of being outside my body, yet in direct relationship with it, was very disturbing to me and created a consciousness of separation and fear that I had not felt before.

At nine years old, I experienced an amplification of that out-of-body reality, laced with profound anxiety and fear. When I lay in bed in the room shared with my sister, I began to spontaneously see certain patterns, such as red and blue concentric circles on a white background. Later I recognized a very similar pattern presented on television stations that had gone off the air. As soon as I began to relax in bed this picture would appear and start moving. While it moved, the liquids in my body—blood, lymph—also moved. (It is known that our bodies are composed mainly of water.) These body fluids seemed to splash back and forth. I had similar sensations later on when I perceived my "light body."

This back and forth sloshing motion created an anxious, nauseated feeling and a sense of spinning that was accompanied by a sound, faster and faster. This unpleasant motor noise droned louder and louder in my ears until I was suddenly sucked into a vortex. I struggled against this pulling, holding tight to the bed, but ultimately the intensity of the sound and the spiraling always peeled me away. I was sucked into a tunnel and then ejected into the sky. At that moment in the sky I felt good again—light, complete, unaware of my body.

After finding myself in the sky, I almost always went to a group of high, dry caves. There were colors on the walls that had a definite palpable structure. There were even smells and all sorts of things my five senses could perceive. It seemed that I would stay there the entire night and then snap back into my body just before awakening in the morning.

I don't remember doing anything there, it was just the feeling of being very contented and above all loved, and so I lingered. I felt whole; it was a warm, safe and dry place.

For a long time I fought against the pattern and the motor noise, which at the onset always terrified me. Without fail, as soon as I had relaxed a little I was swept away in these nightly adventures. I tried leaving the light on, and the radio; I opened the window for fresh air; I fought with my sister. I did everything to avoid sleeping, but in the first offguarded moment that I relaxed, the nightly ritual clicked on.

Sometimes I would end up sitting on the telephone poles. We didn't have street lights then, but there were telephone poles paralleling the houses. As I sat up there looking around, it felt as though I were really physically doing it. One time, showing off to a friend, I boasted, "Tonight I will visit you in your bedroom and prove to you that I was there."

I visited not only him but on other occasions others as well. One night he awoke and was so terrified at seeing me that the following day he told me he didn't want to be my friend anymore. The incident had somehow really shocked him, and it left me feeling more alien than ever.

It took the major part of that year and tremendous effort to finally shut off these nightly excursions. The strength to make it stop came from the overwhelming pain at losing the close connection to my friend. It was the apex of all the negative reactions I had received from the outside world about who I was.

Death experiences and similar transitory states wherein the body is left are experienced in different ways. This happens to more people than we realize. Suddenly one is looking down at himself without warning or knowing how it happened. We can all leave the body; for many of us this openness to other dimensions is somewhat blocked, but being born "open" (unobstructed psychically) isn't as unusual as we may think. Today I know that I was born open, as are many other children. There was definitely a greater expansion of perception that allowed me to consciously experience this psychic openness more intensely than others. Perhaps my mother's blood poisoning during my gestation left me hyperperceptive and conscious enough of other dimensions so that the door to "beyond" remained open. I'm not sure of the real cause. It remains that I was unobstructed, open, and stayed so despite the previously mentioned negative feedback.

The progression into fear states in relation to my awareness of a personal and physical self illuminates the narrowing path into third-dimensional reality. Initially, even though my body was suffering with disease, I had no attachment to it and entered and left without dilemma. By the age of four, when I suddenly "saw" my earth self lying on the operating table, the physical identification to my body self spread panic through my being as I experienced my first sensations of vulnerability: I realized that my fate was tied to that body. The nine-year-old's fearful realization of having no control, coupled with the painful awakening to the laws of causation (if you do this, that will happen), triggered my determination to become as much like everyone else as possible.

However, my early fascination with the thoughts and emotions of others, which I listened to from my own inner telephone system, caused me to blurt out comments about them that I should not have known and left me bereft of an explanation. It is like existing outside of the time zone, responding to thoughts that have not yet been articulated. It actually caused me great embarrassment and amplified that inner angst as to whether or not I was okay.

In seventh grade this "time-interruption" talent led me into a kind of trouble that imprinted me with a gripping fear about the perils of allowing any thoughts or knowings to pop out of me. The introduction of algebra struck some deep chord of recognition within me. Just the sight of algebraic equations would sometimes cause my heart to bound with excitement. I could hardly wait to scribble down the balancing figures on the other side of the equal sign. On our first quiz, I delightedly handed in my paper within just a few minutes of the starting time. The teacher was shocked and immediately suspicious. He called me up and demanded to know how I got the answers. Of course I could only say that they just popped up in my mind's eye as I looked at the equations. The teacher concluded that I was cheating. To be told such a preposterous thing after I had felt so proud of my accomplishment truly broke my heart and created such a block toward math that from then on I could hardly add.

The seemingly impossible task of fitting all my multidimensional experiences into one reality would surely have overwhelmed me had I not been given the grace of someone who understood what was happening and could lift me from the tangled web of these two separate yet overlapping worlds. That someone was Jay Johnson, who invited me to work for her in an exclusive import clothing store. I was fourteen and a half, and for the next four years I spent thirty hours a week in her knowing presence.

She was a psychic who demonstrated all those special talents in a way that made them seem desirable and natural. She encouraged me to acknowledge my intuition, never minding if I preempted time and conversed from a future perspective. She was a magnificent model of multidimensionality, disarmingly charming yet powerfully successful, never proclaiming her specialness but never disavowing her superhuman capabilities.

By the time I was off to college I was actively looking for answers that would dissolve my confusion about these strange realities and allow them to interface with "normal" everyday life. Psychic, telepathic capacities became an intellectual pursuit that I explored in school and throughout my next few years in the Peace Corps in Latin America.

Most of the phenomenological experiences lay deeply submerged in my unconsciousness, intertwined and laden with the feelings of guilt about being "strange" and therefore unacceptable. Fascinatingly, it was, again, the body that provided the bridge between my inner and outer worlds. It was the verifying instrument that produced for me very real physical experiences that validated my inner knowing.

After I moved to Galisteo I founded a bilingual summer school for English- and Spanish-speaking children. Together we discovered how the children's forefathers had lived, what had been special about their generations, their lifestyle and wisdom.

One of our school projects was seeking out the cave dwellings that had earlier been used by the Indians. I had heard that in earlier times Indians lived in caves dug out of the sides of steep cliffs. Fascinated with this depiction, I decided to take the children along on a small expedition to Puye Cliffs, the site of the nearest cliff dwellings. It was intended to be a simple field trip. As we ascended the large canyon and approached the cliffs, I saw something that made me immediately aware that something strange was about to happen and that we had come upon a very special place.

From this distance the otherwise dull ocher-colored rocks were full of tiny holes as though pecked out by birds. The holes had actually been dug out by Indians for logs that were part of their housing structures. Spreading across the entire steep rocky hillside were phosphorescent lichen that seemed to be actually glowing. The rocks were overflowing with these chartreuse lichen. Immediately my entire nervous system reacted. Feeling very drowsy, I simultaneously felt small electric shocks rippling through my body. It was very odd, very physical. I had no idea why I was experiencing these sensations, which seemed to be triggered by the lichen. From our standpoint on the road the caves were not yet visible; we still had to drive around the mesa and climb down over the cliff. As we came up to the top, the physical symptoms increased. We climbed out of the car and followed a small trail, surely one worn down by thousands of footsteps during the previous centuries. We proceeded farther down and over the edge of the cliffs by a small ladder, stopping on a ledge. There, the first cave to my right appeared.

As I entered this cave my nervous system became electrified, as though I'd been shocked. Suddenly, after all these years, I stood at the exact spot I had so frequently visited on those nightly childhood trips. It was like a violent wave of irrefutable recognition. I was totally confused but simultaneously crystal clear. On the inside the cave displayed characteristic markings, colors and smells. My brain urged me to look to the left and notice something. It was already telling me what I would see, as if lifting from my visual memory banks something I had seen just yesterday. Anxiously I glanced to the left and saw the impossible. I recognized markings on a rock with such electrifying certainty that I couldn't shove it away. My brain had previously recorded it, inexplicable as that was. In my everyday life I had never seen pictures of these caves and had never visited such places before. My mind worked like a computer, scanning for details. I knew them all. I had experienced this place before in my nightly astral out-of-body travels. While the children explored around the other caves, I remained in this one, deep in shock.

The long vacuous tunnel of the time continuum collapsed. My consciousness lifted from the attachment to my third-dimensional body, just as it had done so often in childhood. The part of me receiving sensory data was now the same as then; it was timeless, and not aged. Equally, I also had the sensation that my relationship to this cave was not only in the still, lonely nights of the nine-year-old, but also included a flooding of sounds and scenes experienced not only by myself but also by a community of people. Indeed, I had known this cave in an even more extended ripple of reality—a past life.

Later I had a similar multidimensional experience wandering alone in the southern part of New Mexico. I came upon an embankment of flat black rocks, etched with many petroglyphs. Laying my hand on them, I felt suddenly as though I were in another reality, in which I saw with my physical eyes a group of Indians approaching. Again I experienced the amazing flood of information through all my bodily sensory systems while simultaneously being totally unaware of or attached to my own physical body; that is, I could not say whether I was like them or an adult woman of the 1970s. The Indians wore a certain type of loincloth made of pressed-bark fibers and were laden with large baskets on their shoulders, strapped onto their heads; they were foraging for berries. I felt very comfortable, the way one feels among friends. I had no idea how much time actually passed. Suddenly everything disappeared, and I returned into our dimension.

My linear, intellectual mind found it absurd to envision Indians in the middle of the desert picking berries, loading up their baskets. As I hiked farther, I wondered how I had triggered the breakthrough into this other dimension. Perhaps I wasn't well. I paused, held my hand to my head and asked myself what this was and why it had happened to me. Looking ahead, I spotted next to a gnarled bush a sign that read, WILD GRAPES.

I was naturally astounded and wondered if my higher self had intervened at that moment to settle my confusion with this affirmation that indeed there had been Indians who enjoyed the berries and grapes that grew then, in some other time zone.

Later in the day I spoke with the rangers who worked in the area. They confirmed that five hundred years earlier in this region, plentiful vegetation had prevailed: berries, grapes and other edible plants. Perhaps because of severe drought later, the Indians had to leave their cliff-dwelling homes.

This experience provided the impetus for a project with my Galisteo school. The object was to gain knowledge by tracing down impressions and experiences that lay dormant, yet available to us. For example, if we wanted to learn more about the Indian culture, I would take the children along to the petroglyphs engraved on the rock walls of the hogbacks enclosing our valley. I guided them to lay their hands on the markings, close their eyes and simply observe which pictures surfaced before the mind's eye. This produced profound insights, precisely because the children intellectually knew nothing about the Indians, neither which berries they ate nor which textiles they wove.

Gently resting their fingers on the rock markings, the children began to tell stories of what they received. They described the loincloths worn by the Indians. They described the smell of their prepared food and the hunts. I realized that this type of "psychometry" was an astounding technique that could be used to gain knowledge about people, things and

occurrences that had left behind vibratory descriptive patternings. This is "real" history: timeless, multidimensional.

The children developed direct contact with other time
zones, dimensions, places, and they recognized that these were
all real. Real because they felt them through their senses. They
could also interchangeably substantiate details. One child would
say, "I see a man and he's wearing a red loincloth with a zig-
zagging pattern." Another would add to his description, "And
I can see his quiver and arrows," with more details. Later, in
archaeology books, we corroborated the accuracy of their de-
scriptions.

Such experiences never leave the body. It is very important
to recognize that the body can convey such realities and that
they can have profound meaning to us. The body is an inexhaustible teacher. In its own ways it can draw our attention
toward something, identifying it through very physical and
physiological memory packets. The body conceals within itself
experiences and impressions that become revolving information
systems. This "body reality" is so perfect, specific and immediate that it cannot be disputed or rationalized away and
cannot be destroyed.

Years later, returning once more from Latin America, I
felt the pull to Puye Cliffs. As I made my way up the mesa
it was raining. I experienced again the sluggish sleepiness followed by the feeling of being electrified. My body had not let
go of these memories. I felt again how the vibrating lichen
stimulated my electromagnetic field and my memories. I came
along the same way I had sixteen years ago. I wasn't sure if
I would remember where "my" cave was hidden. All at once,
in a flash, the same trust, the same emotional stirrings were
there. I stood before the same cave. My heart beat faster, tears

streamed, I felt the same security that I had coiled up within myself as a nine-year-old. Today I know that in a past life I had found a safe home at that spot. This deeply embedded feeling of snug security had drawn me back. In fact, as I look at my three Puye Cliffs experiences, I realize that the common thread between them all is that in each case I had recently moved and was not yet established in a new home. Thus, my multidimensional consciousness simply zeroed into its strongest available frame of reference for "home." In each case the experience was healing me and allowed me to continue along my life's journey with a heightened sense of centeredness, even though initially I was unable to identify the source for my renewed balance.

SUMMER SOLSTICE IN SOUTH AMERICA

2

Every time we access our multidimensionality, we amplify our capacity to become the manifesters of our lives. Tucked away in these timeless dimensions are levels of knowledge and experience that can dissolve all the blocks that keep us from moving freely and harmoniously within realities of grace and purpose.

The time-continuum format is illusionary and incapable of supporting life because it is of its very nature incomplete and thus dependent on something yet to come in the future as well as itself, created by something already depleted or dead in the past. The present is therefore tenuous and uncelebrated, as the consciousness is too preoccupied with the rigors of passage. Perception, then, falls prey to the profoundly destructive view of life as struggle, resistance, control and necessary limitation, rather than reality as constant change, unveiling wonderment and endless resource.

The entrance to other dimensions is usually revealed during unusual circumstances in which we "unintentionally" become involved; the experience available to us by contacting these dimensions merits learning to integrate them skillfully into our daily lives. Illness and trauma are all too often the triggers that force the body to find a bridge to its source, the soul. Normally, we are separated from direct contact with the spiritual dimension, but when the body is endangered, we are compelled to experience holographic (all-encompassing, multidimensional) beingness. During my early traveling, I was unaware that these adventures were interfacing with each other in a way that would later become the foundation of my work. These points of data exchange were always clusters of multidimensional experiences that altered and enhanced reality. Perhaps the healing theme was specific to my own growth and, because of its dramatic nature, provided for me a tangible

imprint of its effectiveness as a life tool. All the healing techniques I encountered were multidimensional in nature and
drew upon those specific laws of interfacing worlds (i.e., spiritual and physical) to produce what were truly miraculous
results. Here, then, were the seeds of the guiding principles
for the future Light Institute.

I went to Mexico to study a few semesters of premed. I
lived there in a small pueblo named Contadero, a one-street
town. It is situated deep within the national forest above
Mexico City. The inhabitants brewed pulque, an offensively
strong-smelling alcoholic beverage from the cactus plant. There
I was, a young foreign student, living in the hills far removed
from Mexico City, where I studied. I had rented a small house
and was taken care of by Isabelita, an older woman. Whenever
I had stomachaches or fever she always gave me certain herbs.
She described the healing powers of the different plants and
began to instruct me regularly. I recognized more and more
that within all living things were certain active properties.
Perhaps in one plant a cheerful spirit was active, in another
a dominating one. According to the type of plant used, one
would absorb that spirit which would confront the spirit in
his own body, setting the healing process in motion. As I
became aware of such a reality, I realized that things happened
in the world which couldn't be scientifically explained, but
were nonetheless real.

Isabelita was a Curandera, a healer of infinite wisdom,
although she herself never formally assumed this title. It was
her idea to lead me craftily, albeit with a well-intentioned
spiritual purpose, into a magical world differing greatly from
the reputable, dry and flat space-time continuum of the university. She taught me to appreciate that reality can occur

even if it isn't controlled by our analytically logical conception of the world. Such extraordinary occurrences can be very useful and meaningful to us all.

She brought me closer again to animals. I had communed with animals on spiritual levels as a child, but she was the first person in my adult life who acknowledged this pathway to nature and who didn't avoid showing it to me. It was a breathtaking existence for me—Isabelita and the medical environment at the university. I was an excellent student. Fate intervened, however, and I prematurely returned home when Mother became very ill, needing heart surgery. By the time her condition improved, I had already become fascinated with the work of the Peace Corps and decided to join the first waves of volunteers in Central America and South America. Learning Spanish in Mexico opened up an entirely new vista of myself as a person. My English self was totally serious and profound, yet my Spanish self was delightfully jovial and cordial.

Europeans consider it normal to speak a second language. As an American, it came as a wonderful surprise to me to discover that in using a different language, I could experience new characteristics. It was exciting to realize that in English I thought and spoke very profoundly, earnestly. Nothing escaped my attention that was not examined for the inner meaning. In short, I was filled with the seriousness of life. When I spoke Spanish, however, I was a comparatively high-spirited, laughing being, whom the Mexicans called "la graciosa" (charming clown). A whole new part of me had been born.

The Peace Corps enhanced the daring explorer in me. I could hardly believe that someone would teach me a new language, send me into a foreign country, give me the opportunity to discover a new culture, and pay me for it. This was an unbelievable adventure. Before long I was on my way to El Salvador, in Central America.

I lived and worked in a small town named Dulce Nombre de Maria, especially auspicious since my name is Christiana Maria. This town, near the border of Honduras and Salvador, became a magical place for me. I was filled with a sense of "untouchability," a quality peculiar to youth. Here in this small foreign town I was to offer help on hygiene, child care and nutrition. It was wonderful to have so much responsibility and to feel so purposeful.

This situation brought out the multidimensional person in me: someone capable of guiding and healing and who could accomplish things that I, as a young American, had not dreamed of doing. I truly forgot my boundaries. The people expected me to know and do the impossible, and therefore I did.

I communicated with them on a "magical" level. I felt magical, the people thought I was magical and I thought they were too. We interacted with one another in a way I'd never experienced before. As a child, I had always protected my magical world from any outer-world intrusion. Here I could openly express my inner experiences without continually being asked, "Why?"

Unbeknownst to me, this was also to be a first course in the limitations of personal power. My sheer bravado invited trouble. I had rented a house belonging to a woman who always quarreled with her husband. When the fighting was extreme she would evict her tenant and move herself in. At one point she wanted to move me out, but I wouldn't budge. Her response was to engage the best sorceress in the Salvador area to dispose of me, by death if necessary.

All my friends and neighbors were very worried about me and said, "Niña Christina, you must move, do something, but don't irritate this woman." I held my ground. So the landlady contacted the sorceress, setting in motion a set of remarkable occurrences.

The people told me a lot about their mythology, about sorcery, *brujeria,* how people could be controlled or even killed, but I always insisted that "white magic" could hold sway over any negativity.

It was like an initiation for me and became very important in my later work. What we accept as real and true to the utmost limits of our sensitivity and understanding is really only the tip of the iceberg. My limited vantage point assured me that I was outside the sorceress's scope of influence after all, a member of the "material" world. Despite this conscious denial, I was pulled into the initiation, into a deeper level of understanding. I became deathly ill with malaria, which was naturally explained by everybody as the *bruja*'s curse. Today I realize that I willingly agreed to serve as an instrument in her magic, as my own initiation. Naturally, it was not possible for her to kill me, and because I continued to live, I became something of a mythical being in this little town. Until then the *bruja* had been successful with all magical intentions, even purportedly killing more than a few people.

My malaria attacks were accompanied by high fever, which created an eerie dance between the sensation of being "out of my head" and a kind of lucid awareness that included glimpses into the future. During one of these attacks, I saw a large moving blackness fall upon our town and move through it. When my attack subsided I told several friends. A few weeks later the entire town was overrun by army ants, a black, shapeless, vibrating mass moving itself across the town devouring everything. Thus we had a subsequent explanation for my vision.

Myths or beliefs are called *creancias* in Spanish. They are stories that people have told from the beginning of time. Some of the myths are of a spiritual nature. In Latin America, as in all "native" cultures, spirit and matter are not considered separate; the body is seen as a sort of megaphone for the spirit and carries out the dance of the spirit—"whole in the spirit, whole in the body." When illnesses are treated, they may not even treat the body, but only the spirit as the source of the disease. The native people told me stories about creation, about the first inhabitation of their country, and gave me explanations about things now happening in our bodies. They told me all their history, mythology and fairy tales. Since I was very left-brain oriented at that time, I noted all the stories carefully. They had passed down these stories generation to generation from the beginning of their existence, never losing any of the dramatic energy in the stories. While I was busy with my documentation, I noticed that in their reactions to the world around them, these myths weren't "past tense" to them but "present tense."

One couldn't be sure if the "bogeyman," appearing in his menacing mask before someone's door, had terrified the family last night, or if he had appeared before their late great-grandmother. The people had allowed time to become one large sea, an uninterrupted pulsating flow of life, in which the here and now had no effect on the power of experiences to which the body was subjected. There was incredible force in the emotional content of their storytelling. Other realities and forces, which appeared to direct the lives of these people, began to fascinate me. They distrusted the modern medical system and its knowledge of bacteria and viruses. They felt that people lived or died according to mythologically based influences. Even science has no explanation for why one becomes sick while another remains healthy, although both were exposed to the same bacteria or viruses. I began to observe that health was directly related to our emotional interpretation of ourselves. If one had a strong sense of oneself, there was less etiology of disease; if not, and one identified more strongly with outside forces, there was a much higher incidence of disease including symptoms of suspected psychic origin.

Similar processes of psychic identification are universal, especially visible in children. One reads a book or sees a movie and is so engrossed in it, perhaps identifying with one or another of the characters portrayed (at least for the duration of the book or movie), that one perceives the story as reality. These processes are more meaningful than is commonly accepted. They portray one of any experiential situations in which an impression from a mythical experience can be just as effective as a "real" experience, played out in the 3-D world of time and space. Through these experience-identification processes, we have a starting point in the exploration of the astral dimension, in which the sea of consciousness is expanded. Although the existence of levels of consciousness in astral regions may not be clear, they nevertheless have a very subtle and simultaneously enduring effect on us—whether this consciousness is manifested through history and stories of our family, clan or group, or through other layers of consciousness within ourselves. The energetics of the astral dimension are emotional in nature and can therefore access our emotional body from a nondiscriminating level that is often unconscious and therefore capable of impregnating us with experiential realities that may not be those we would consciously choose ourselves.

Thus, El Salvador was a challenging tool for exploration into the astral dimension and helped me to begin the vital perception of that dimension's laws of existence. I was very careful then about "scientifically" collecting and documenting data, to "prove" real every supernatural occurrence. Later I read through these notes and realized that they weren't important to me, as I knew experientially what was true. Upon my departure I discarded them, the way I had discarded my books in moving from one house to another in Galisteo. In both cases my Higher Self made clear to me, "You are the book."

After we had lived awhile in Bolivia we moved on to Paraguay. This offered me the opportunity to expand my human expression further by learning the native language of Guaraní. It is the second official language of Paraguay, and they are very proud of it. Many Paraguayans sense that it is the language of their soul and put it before Spanish. They aren't ashamed or embarrassed of their Indian heritage, nor do they feel illiterate speaking Guaraní. They consider their own language to be very rich and lively and better suited to express their inner self than the foreign tongue of Spanish.

I was one of the few foreigners who had ever learned Guaraní. I was able to change over from one form of consciousness into another, to transfer myself out of my own reality into the realm of their reality. When the word spread that I was interested in hearing their mythologies and tales of the supernatural, Paraguayans began coming from all over the country to visit me and relate their experiences. I didn't know then why they visited me. Today I would say that it was naturally my karma, my destiny, to be at this time and place with them.

Just telling me their stories seemed to have a powerful effect on them. Possibly they saw me as a type of shaman. They felt as good in my presence as I did in theirs. Our connection was heart to heart. I spoke with them about plants. I used my herbs to strengthen them, so that they could find the healing power in themselves.

This is not what they were used to hearing from outsiders. They considered me to be someone who bridged their souls, not to their everyday outer world but to their inner world. This was why we always conversed in Guaraní, although many could have expressed themselves in Spanish.

While in Paraguay I experienced something very profound, which influenced my life considerably. Everywhere in South America the summer solstice is celebrated in special ways. This longest day of the year becomes the focal point of joy and the spiritual elements to life. People gather, laugh, dance and light bonfires. In many other cultures, as in the early days of the Celtics, for example, the celebration of summer solstice had a deep spiritual meaning as well.

I employed a twenty-two-year-old woman to help out with my small children and the household. Martina helped me to become fluent in Guaraní. A few days before the summer solstice she invited me to attend a special celebration of this great day of life force. Since we were friends and I was always interested in customs and rituals, I went along. I had no real idea of what was to take place.

We drove by bus to an outlying suburb and saw the wood being stacked for a giant bonfire. I had already seen a similar bonfire at summer solstice celebrations in Bolivia and assumed that it served as a symbolic energy cleanser, portraying the life force. While we passed the afternoon chatting, the fire had been lit, grew to a blaze, and subsequently burned down to glowing embers. I hadn't paid much attention to how they had raked out a long path in the still red-glowing cinders. People stood around this path, the glowing embers were spread out in a thick blanket, and suddenly my friend walked straight over the fire.

I was speechless. This was a completely normal young woman, with a child of her own, who had never displayed any leaning toward the mystical. We had often conversed very intimately and we were very much alike. Yet she simply walked over a thirteen-foot-long path of hot glowing coals. Many of the others followed her!

Back in 1967, firewalking wasn't as well known as it is today. Now we know that we can learn a lot from firewalking. None of this applied to the young woman who simply walked across because "one renews oneself this way on the summer solstice." Later I learned that she had been accustomed to this since childhood.

Witnessing this ceremony forced my consciousness to break out of its own limitations. I couldn't cope with the shock of seeing her walk across unceremoniously, without prior fasting or any spiritual preparation. I had heard of firewalking before, as part of religious rituals, but not in this simple, unpretentious fashion. I was struck again by the spirit of unlimitedness existing within us, despite our usual separation from it.

Since then I've walked over glowing coals twenty times, and so have five of my six children. Since I've done it in different places and at different times, I know that there's no "trick" involved. How hot the coals are plays no part. The coals are simply hot, and when we walk over them we don't know how it is possible. We are just beginning to understand that the body is a part of our broad multidimensional beingness and not the limited vehicle it is usually considered to be. The Paraguayans had no explanation for how they could do it— it was considered very ordinary.

My daughter Megan was twelve when she first firewalked. I was delighted to watch how she stood first in line to walk fearlessly across, while the rest of us looked on hesitantly. She didn't wait. She dashed to the edge, took a deep breath, paused shortly, and proceeded across. Afterward she said to me, "Now I know that I'll never get cancer, maybe I'll never even die." Her brilliant cognition struck me like lightning.

Megan simply surrendered her self-will without fight or opposition to prove to herself that she could naturally firewalk. An interesting aspect is that this incredible firewalking initiation becomes easier when one sees someone else do it first. The brain registers it as possible, and thus it is.

This insight is important to my current work. I help people to let their visualizations expand their reality. Whether they see others fly or pass through walls or see the incurably ill cured, if the mind perceives something as being possible, it will develop this capability in itself.

In my work with children I've accumulated numerous experiences proving that children can accomplish things we adults consider miraculous, as long as they don't accept limitations that convince them it's impossible.

A wonderful example is an incident from my summer school in which the kids had completed a Silva Mind Control pilot project for children. They were instructed how to stop bleeding by laying their hands on the wound, among other healing techniques. The kids were playing hide-and-seek and one of them cut her leg deeply on some barbed wire. Even though I wasn't around, the other children responded immediately, laying their hands over the leg, visualizing that they would stop the bleeding this way. Within minutes the bleeding, which had been gushing down her leg, stopped. The girl was left with a scar, perhaps because they forgot to visualize also that a scar shouldn't be allowed to form; but it was a breathtaking accomplishment.

This incident was another confirmation that the spirit can rise above adversity and be effective as long as the self-imposed handicaps and limitations cease.

Our heart isn't afraid of the unknown: it fears only that which it has already classified or known. When we don't receive the impulse, "Mama said to be careful, this is dangerous," then we also won't encounter the normal limitations.

Most remarkable during my work in Central America and South America was witnessing the miracles that occurred among these insular people, living as generations before had lived, in the same place, never venturing outward. Impossible things happened. I was always confronted with wonders and was delicately guided by their example in consciously and willingly moving into those levels and situations that release the self-imposed limits of daily life. I clearly felt that there was much more than what usually meets the eye. I realized that the connection between any outer- and inner-world occurrences can be found on a spiritual level, which has its own rules of motion. Where the unmanifest formlessness penetrates the reality of our daily third-dimensional lives and is accepted as reality by becoming a part of humanity, of the family, the culture, then life becomes truly a magnificent adventure.

It makes no difference whether we deny or accept these other reality potentials. They exist. Myths and their experiential phenomena activate their own effective energy. I was just beginning to understand how energy flows between formlessness and our material world of forms. We can tap this energy within us and consciously perceive it influencing the reality of our day-to-day lives. I was fascinated with experiencing such energies. I wanted to discover more and more of them and learn to use them directly to influence health the way I had witnessed them doing in South America. These profound teachings created the structural components of my work, namely to expand ourselves past the limiting confines of the time continuum, allowing us to access convergent multidimensional energies that then amplify our cognition of ourselves as part of a cosmic, direct whole.

People coming to the Light Institute to widen their spiritual horizons perceive these energies and learn to integrate them.

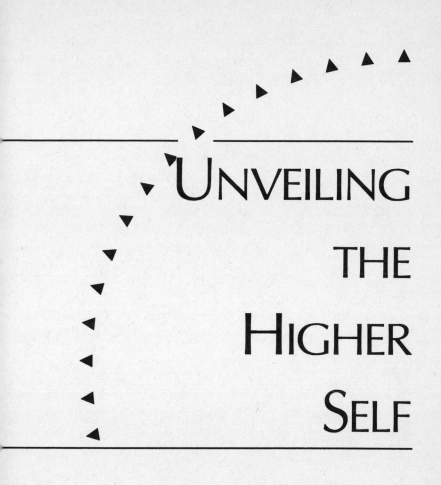

UNVEILING THE HIGHER SELF

3

Insight doesn't unfold in a linear fashion, but more often holographically, on several different levels at once.

I had been sent to Latin America to deal mainly on the physical body level, but it became fascinatingly apparent to me that the physical body tells the story of the spiritual body. Any knowing that brings with it a deep sense of truth is "felt" as a resonant chord throughout our whole being. We actually have four vehicles of consciousness that converge to orchestrate reality. These are our physical, emotional, mental and spiritual bodies that are the actual components, or perceptions, of life. They all interweave and influence one another in the most subtle and remarkable ways. However, it is the spiritual body that is the ultimate vehicle of transcendence for them all. The spiritual body interfaces with the soul and therefore influences the other bodies to respond and search out those whisperings that access our divine aspects—even though we are often completely unconscious of the goings-on. The soul is, of course, part of the universal God-force that on the individual level is mirrored by what I have come to call the "Higher Self." It is the Higher Self that extends down into manifest realms through thought and light. We are born accompanied by the grace of the Higher Self, which is not a personage but an energetic channel of the unmanifest soul. As we grow, our focal point of attention and awareness is drawn down from those higher octaves into the physical, material realities where we lose remembrance of our most divine relationship with the Higher Self. Though at any moment we could unveil and again re-connect with the Higher Self, we usually only do so when we are somehow endangered in the third dimension and extraordinary experiences motivate us to grasp that emerging link in order to actually survive.

During my duties in sick wards I discovered that massaging

or pressing the body in certain spots would trigger sudden memories out of the past, a past that often depicted totally different environments from earlier lifetimes. Children talked about earlier lives, other bodies they occupied. They reported very distinct physical feelings and reactions in their current bodies that obviously developed in other lifetimes. I realized that the physical body carries imprinted in itself information of who we are on the highest ethereal levels. All of this prompted me to explore the possibility of deep-reaching energetic processes, wherein the more subtle emotional and spiritual vehicles could have a voice.

Upon my return to New Mexico I learned connective-tissue massage and began teaching esoteric massage at the College of Natural Medicine in Santa Fe. At that time a school for acupuncture was also founded at the college. Since many of my clients said they felt acupuncture points and meridians were activated when I carried out energetic processes with them, I decided to take up this new field.

I had already felt the imbalance in patient-doctor relationships, not only during my service with the Peace Corps but also in American well-baby clinics where I worked. The doctor holds the power and authority, while patients are never asked their opinions of their illnesses. After witnessing the interaction of the subtle bodies (emotional, mental and spiritual) in disease, it seemed ludicrous to me to pretend that the patient was powerless when in fact I could see how he was participating in and creating his disease through the energetics of the subtle bodies. If he could not be made aware of the messages he had received from his multidimensional self, either he didn't heal or he developed a new series of symptoms in the body that described those same messages or themes in a different way.

When the time came for me to stick needles into people

I found myself resisting. It appeared in no way better than or different from allopathic medicine; this style of acupuncture was routine Western medicine to me. I was supposed to suddenly assume authority, and with the needles at my disposal, poke them into others and pretend that I, not the patient, was the one doing something. Was I now to refute my own experiences, act against my own feelings and pose as an external authority to take power away from others?

I decided I had had enough of acupuncture. I had learned the principles of the meridians and energy flows in the body. That was very helpful for my subtle body work. I wanted no more. But on the day of this decision something remarkable happened. Since I intended not to continue studying acupuncture, I wasn't fully concentrating in class during the lecture. Suddenly and seemingly out-of-the-blue, the teacher began to lecture on "windows to the sky." I reacted to this expression as though my body had been electrocuted, the hair on my neck stood up stiff, my heart pounded and I began panting. I sat in a state of full alert as he explained that there were esoteric points that for centuries and millennia had been kept secret. They were used only to help the spiritually elite in establishing contact with their inner selves. These windows are connecting points to higher octaves of consciousness and are the actual thresholds or merging points for the subtle bodies as they interface with the physical, earthly body.

After class I spoke extensively with the teacher, who confessed he couldn't explain, that he had had no intention of discussing this esoteric topic except for a sudden impulse. Again I experienced the pounding heart and the physiological confirmation that this had been for me! In the following days and nights, I received information through many spontaneous remembrances and powerful images flooding through me. I actually experienced myself from another dimension, out of body, transmitting the "windows to the sky" points to those on the earthly plane.

From then on I saw acupuncture in another light. I studied and experimented sometimes fifteen hours a day in order to crystallize my memory to a clear focus. I was always guided to explore new esoteric points, their potentials and methods of use and their connections with the other points. It was indeed the patterns created by linking up certain points that motivated specific predictable results. In Chinese acupuncture, the sky embodies the spirit. It is a larger, broader range of reality, belonging as much to us as does daily life, although these octaves of consciousness usually remain unfamiliar to "everyman."

This period of learning determined what has now become the foundation of our work at the Light Institute. It provided the techniques that open up new aspects of one's multidimensional self, with everyday life always taking on new perspectives as a result.

The interaction of the four bodies became more visible as I explored consciousness from the holographic perspective. Processes change dramatically in nature the moment we pierce the veil of the linear, intellectual mental-body. To do this, we must move from the outer beta brain frequencies to the slower alpha and theta range. Then, as the mind is refocused, it is able to see cause and effect spinning together in a creative dance. What I call the "higher mind" comes into play and the cosmic language of image and color explores new avenues of meaning.

The higher mind has the holographic capability of simultaneously recognizing and grasping information from many dimensions, which are directly related to our own personal repertoire. This is an important point: We are not set loose in other dimensions like moviegoers to passively watch something that is not directly related to ourselves. WE are the movie. What we experience is about US. The Higher Mind always uses its own, albeit expanded, repertoire of cognition. It is really the birthright of everyone to make use of his and her full-mind capacities.

The spiritual body portrays the highest ethereal aspect of our beingness. It is the essence of our godly selves, the source of our being, the outline of our lifeplan. It impels us out of a formless condition of pure spiritual energy and coalesces us into the material realm of our body, the soul, that godly source in us that has no particular form. Our spiritual body is an echo of that soul. When I speak of the spiritual body, I normally mean a beingness harbored by the soul, although the soul itself has no form. During this time I also gained a clearer understanding of body anatomy, including the astral body, which is almost unnoticeable in everyday life. The astral body is actually the active instrument of the emotional body, which is the most fascinating of all the bodies.

In the great evolution of the "two-leggeds," as the Indians named homo sapiens, the emotional body has remained decidedly behind the rest of our bodies. It is a highly complicated and astounding essence that lives, so to speak, within us and internally dictates the quality and substance of interchange among all the bodies.

With the passing of the ages our physical bodies have become stronger, live longer, and are healthier than ever before in the history of mankind. Our mental bodies have developed even further, we stride forward in the development of inner and outer worlds, our intellectual knowledge broadens. Unfortunately, it has gone so far that we have isolated ourselves almost entirely within our mental bodies to the point of damaging our wholeness. In other words, we sooner accept truth from the level of reason than from the heart, or other universal octaves.

There are, of course, some "side-alliances" that are very important for us to consider so we can maximize their effect on us. For example, our mental body can actually control the physical body! Through research in the last twenty years we have discovered that attitude and the mind can control the physical body. Some can willfully reduce their heartbeat, others can influence their biochemical discharges. This discovery was how research into biofeedback phenomena began. In the meantime there is a lot of data proving that voluntary as well as involuntary bodily functions are controllable in this way, through the avenue of the mind. Additionally, we have indications of positive influence on cancer with biofeedback and other mental techniques.

Important: Attitude and understanding do not control the emotional body! To the contrary, it is the emotional body that determines our existence on all levels of consciousness on this planet, and yet it has remained behind in its own conscious development. The reason for this is that emotionality on energetic levels belongs to the astral dimension, which is decidedly outside time-influenced reality. Unaware that "time passes," the emotional body continually revisits or reassembles emotional components of its own design. Since we ourselves have become so identified with our mental body, we practice the illusion that we are influencing and directing the emotional body with our conscious will. For example, we might desire a behavioral change so that we do not experience anger. Naturally, we can cover up the anger with a practiced attitude, hiding its outer visibility so that we don't live under its thumb or divulge it to others. However, when it is repressed in this way, it will emerge, at some point, in another way—in the form of illness, for example, or by constantly finding ourselves witnessing other angry people, who are in fact acting out our anger for us.

Since the consciousness of any of our bodies is not dependent on the actual material vehicle, our "sticky" imprints from the emotional body just revisit themselves through each incarnation. The "old" emotional body brings into the "new" physical body all those experiences, reactions, and perceptions of reality that it gained in other bodies.

The plan of the spiritual body always begins with the intent to embody itself physically for the purpose of gathering experience for the soul's development. The emotional body orchestrates those choices from an experiential perspective and so often creates a "Tiger Chasing Its Tail" amplification of the themes the soul is exploring. This is exactly what is happening to us in the third dimension. We are so narrowly focused on our negative themes that we cannot believe we are masters! We might experience fear in certain situations, such as seeing snakes biting in violent arguments, for example, and this is biochemically and electromagnetically stored within us. One is not easily rid of this. Fear creates this astral charge in an energy form filled with the thoughts, feelings, guilt and choices that fear, worry and upsets bring along. All this creates a fine imperceptible net, unhampered by time and space. Whenever this unconscious energy is restimulated, these emotions, thoughts or reactions are always reactivated.

To pose a difficult example: What happens if we kill another person? It is against divine law to kill, but we do it nevertheless, perhaps to gain a definite teaching about life and death. The truth is that nothing happens outside the choice of our soul. In terms of our spiritual development we must learn to understand not only the perpetual cycle of life and death but also that death is only a passage. We may take the life of another, but the "cosmic law of permission" is always in play. To live within these divine octaves we must understand that any action that makes a choice for another is outside of balance, whether that is picking a plant, killing an animal or one of our own kind. No being would allow another to act for him or her, especially to dominate his life—except to offer a profound gift of spiritual proportion. However, because our self-judgment and the impressions of such an act are so intense and penetrating to our emotional body, we are almost never able to align it with the soul's life plan, which may assert, "It serves a purpose in teaching that the victim and the victimizer are one: both have chosen these roles to participate in the experiment of life."

These traumas are biochemically and electromagnetically stored within the matrix of the emotional body. The imprint is carried along like an electrical impulse by the emotional body into each incarnation. We come into a new life with a certain life plan demanding certain experiences for fulfillment. These experiences, however, will always be met by the repertoire of emotional reactions already stored by the emotional body. The emotional body can't differentiate past, present and future. Through its storehouse of astral energy, it magnetically attracts people and conditions that will verify those experiences within its frame of reference. This is not a continuum, but rather an outward expanding arc like a ripple reverberating in a pond.

It is almost impossible to break free of this cyclic treadmill, since the emotional body is bonded to the physical body through the solar plexus center. This solar plexus chakra stimulates the sympathetic nervous system, which works on the flight-or-fight principle. If one experiences something frightening, the ganglia in the solar plexus are stimulated, activating a change in the chemical composition of the blood in the brain. This in turn creates a form of electrical stimulation to which our emotional and physical bodies become addicted with each recurring episode. Our own conscious emotional body searches for those people and conditions that provoked this electrical stimulation, so that they can be reenacted and again experienced. This pattern becomes unnoticeable with time, and we cease to experience it directly, yet continue to perpetrate the cycle.

In contrast, small children indicate high sensitivity, often reflexively, when they sense aggression. For example, they detect when we unconsciously raise our voices a bit, and they are very sensitive to our body language. If they possess a repertoire of the slower vibrating emotions such as fear or anger, they grasp immediately and intuitively which of these energies is being expressed through our tone of voice or body language. The children begin then to adapt to those energies and emotional reactions germane to themselves, and to seek them again and again. Even if one doesn't embrace the concept of reincarnation, it suffices to observe small children in such situations and see how the impressions they receive are so reinforced that they practically become second nature to them. This process of conscious or unconscious influencing is commonly aided by adults who heighten and reinforce the general emotional imprints with specific emotionally charged words, such as "Be careful of" or "Watch out for" or "Stay away from that person."

In this manner impressions endure, influencing the ego

from birth onward, and becoming part of the physiological functions. These reactions will occur outside of the appropriate judgment of time and space. We frequently observe forty-year-old or even sixty-year-old people who still answer to their parents as though they were only twenty or younger.

These involuntary reactions are brought forth by the astral energy in the emotional body. The astral level is like a mirror image of the physical level. It gives the emotional body a subtle structure in which to operate. If we have out-of-body experiences, for example, as I did as a child, we see ourselves there as a duplicate of our physical body. This duplicate actually "travels" and perceives other realities, such as the astral dimension, from emotionally determined estimations.

In this light it should be mentioned that "ghosts" are astral in nature and should not be confused with souls or the spiritual body. There are ghost beings that float through astral dimensions and are fascinated with certain places because of their emotional attachments, expressing themselves for long periods only as astral energy remnants of earthly incarnations.

My spiritual background was first determined by psychology, then I became acquainted with mythologies of man, followed by my work with the physical body. Thereby I discovered that the mental body of man could not control the emotional body. We naturally wear masks that can conceal the emotional body, but we can't change its true effect upon our lives. It is possible to influence our behavior, but on other levels we are still subjected to old emotional patterns until we learn how to clear them.

I observed that people constantly reinvented the same experiences. For example, if a woman had an unresolved relationship to her father, she would search for a husband displaying similar characteristics and behaviors. After a certain time she would experience herself stuck again in the same old relationship problem, would divorce, and then marry another man. After a time she'd again recognize the same problem confronting her, although perhaps cloaked differently.

I have also learned that even with the best intentions it is impossible to change the reactivity of the emotional body by conscious decisions. From my experience in Latin America I knew, however, that through a conscious connection to spiritual energy the emotional body could be influenced.

Up until this point in the history of this planet, Earth has generally been used as a place for physical development, a "laboratory for genetic experiments." Until now we haven't realized that Earth also offers development possibilities for our emotional body, which needn't only store anger, fear, isolation, guilt, but might include such feelings as ecstasy, rapture or bliss. Our vocabulary is poorly equipped to describe such states as these lofty emotional octaves, because we simply don't experience them! We refer to happiness or joy, but these are only concepts—not fully experienced.

I would say that in the evolution of the emotional body the orgasm comes closest to the higher octaves of rapidly vibrating emotions. The orgasmic condition closely resembles ecstasy and rapture, as far as most people have experienced. How long can we maintain this state, how long can we remain in the vibrating frequency of orgasm? A half minute? Perhaps. And then we immobilize the frequency again and shut ourselves off.

We become electrically so charged that we are afraid to lose ourselves. If we could consciously remain in this frequency, we would give up our separation and actually merge. Because of our fear, we break off the moment of this energy because we can't withstand high, rapidly vibrating frequencies. If we could stay with it, we'd experience an energy potential that would allow us to pass through walls.

For many years, as I observed people in other dimensions of consciousness, I thought that we needed a new vocabulary to express these new things. Later I realized that it would be premature because we hardly experience these vibrations of ecstasy, rapture and bliss at all. At best we think of them. There are only a few people who know what I mean when I mention ecstatic states. Outside of orgasmic physiological stimulation, some people experience ecstatic spiritual states through meditation. We've heard about saints and yogis who float in ecstasy (e.g., Maharishi Mahesh Yogi), go through walls (e.g., St. Theresa, St. Mary, Tibetan lamas), or break through to unlimited knowledge of self (e.g., Yogananda, Muktananda), freeing themselves from the otherwise dominating slow vibration of the ego. If they achieve the great oneness with the universal energy—representing ecstasy or rapture—then they reach a consciousness that departs from separateness via the physical body.

In general we don't experience anything similar in our everyday lives. It poses a challenge to us; it is our destiny and our purpose in life to learn, now, to create the bridge to other dimensions so that all of us experience these levels of reality. We must evolve the emotional body to a new octave. Since the emotional body doesn't recognize time and space, and continually lives by its old habit patterns, how can one achieve this?

I found out that effective alterations in the emotional body, in which fear, anger and all other experienced emotional reactions were released, could only be produced through experiences on spiritual levels. If one comes into direct contact with his god self, establishing contact with his higher self, this creates an acceleration of the vibrating frequency of the emotional body, and thereby releases those burdensome lower energies. If one consciously opens his spiritual body and allows experiences in spiritual dimensions, transformations of mood, insight, feelings and reactions are set in motion.

The work with the acupuncture "windows to the sky" brought forth the first concrete results. These spiritually operative points permit the unmanifest to "rain down," to manifest. The formless spiritual body can consciously establish contact with the physical and emotional bodies. When it does, a powerful energy explosion is attained, so drastically accelerated in vibration that a frequency alteration results. It is this process alone that is capable of changing the emotional body.

In other words, it is through the grace of the spiritual body that the emotional body receives a new frame of reference, that it recognizes new patterns, a new reality with which it can align. Once the emotional body perceives and adapts to this new reality it is then permanently changed.

So began my work. Through these resonances the clients entered altered states of consciousness; they experienced ecstatic feelings so that they forgot that they were sick, that their body hurt, that they hated themselves, or that they worried futilely over something. They forgot their isolation and experienced these wonderful realities that changed them, that imprinted a new frequency in the emotional body, and that they would always endeavor to bring into reality again and again.

A definite understanding of the emotional body is really the foundation of the work. Once we begin to see the emotional body as an instrument of change and a living entity, instead of just a collection of feelings, it is more accessible to change. In fact, *only* then can it be changed. Only when we consciously glimpse into the repertoire of its embedded reaction patterns, into thought forms on the astral level, into other forms of life, through which our emotional body has previously expressed itself, can we create the determination and confidence to newly "program" the emotional body.

I learned to palpate the emotional body, in other words, to massage it, fondle it, embrace it, and fully identify it. This can only occur through the connection with the spiritual body.

In our sessions we enable the client to come into contact with his own emotional body, to acknowledge it, to connect with it, to see it in a positive light. Thereafter we can proceed into new depths, finding a whole world of knowledge, experience and reality with meaning and purpose.

This is the greatest adventure of every living being. As humans, we are privileged to travel through our consciousness into points of light, orbits of synchronistic reality to choose our birth, choose our death, to release, refine and forge aspects of our selves, to change the past, mold the future and bask in the multidimensional totality of our divine nature.

In our first session at the Light Institute we give the client a basic awareness of the emotional body's functions. We explain how experience can be gained in the center of the solar plexus and that the emotional body anchors itself to this spot. The client understands this initial description of the emotional body only when he or she experiences for him- or herself its subjective existence and its deeply anchored emotional personalities. At this point some of the emotional body's mechanisms become clear; its energy and sculpting become perceptible and the client feels a great surge of power and delight at being in concert with his being.

Access to the emotional body occurs in an atmosphere of peace and letting go of burdens. The client is able to lift away a large portion of the emotional treadmills, such as guilt or negativity, and begins to see the self in a new light of love.

Work with the emotional body brings along with it a spontaneous opening to spiritual dimensions. This is really a precursor to conscious awareness of the Higher Self, which is actually the purpose of our work. If we ask ourselves, "What would my Higher Self look like?" then we proceed along our search as though we imagined ourselves gods. We have such linear styles of imagination that we always think of a *form* in this light, of a human figure. We identify ourselves as humans from the standpoint of our ego selves. This estimate is naturally not the entire truth because only a small part of our multi-dimensional soul is expressed in our human form.

It is such a wonderful moment when the Higher Self is asked to take form, and we surrender our expectations to that space that breathes, when the Higher Self so exquisitely speaks to us in the language of form, using a framework of symbols. It assumes the form that our heart and mind can identify. It is the first rung of divine influence in the limitless sea of cosmic ripples. From time to time, or even within sessions, these symbols change. We may feel ecstasy and rapture in the presence of a cube, cloud, hill or certain color. The symbols portray aspects of the Higher Self and contain a vibration that fills the heart and allows our emotional body access to higher octaves.

Pictures, symbols, sound and color form a common matrix of communicating language of our unconscious mind and our multidimensional realities. We begin to learn this language during the sessions. The Higher Self is thus able to ensure that the ego and the finite mind can make the leap into expanded awareness with a level of corresponding integration. The Higher Self's selection of "multimedia" props is a breathtaking display of all-knowing. It actually creates an environment and a sense of recognition which allows us to surrender on levels we've never reached in our entire lives. As the symbols of the Higher Self change in order to address other areas of our beingness, it is as if we are always discovering a new facet of our self.

To illustrate the point: Perhaps initially Higher Self comes in as a figure cloaked in a brown robe. We ask the Higher Self to take us into a scenario that needs to be changed. Suddenly we are in a hall with a group of brown-robed monks. It is a lifetime of austere existence and stern authority. We follow all the rules that are supposedly set down to assure divine presence in our life. But we feel none and waste away, embittered and shut off from any meaningful purpose. We die angry at God's inaccessibility and man's stupidity.

As we release this lifetime we are immediately struck with its correlation to our present life in which from childhood we ardently resisted our parents' attempts to have us attend church every week, and we find that we are still stuck in a monklike existence in which we hide our fear of relationships behind a mist of distance. Coming into contact with the sound of that distance, we experience a time opening in our heart—a sense of soft vulnerability that floods us with hope. As the session closes, we see the higher self as a glowing bright light all around us.

The Higher Self does not always use its own image to create the necessary associations, but its presence sets the frequency range in which we must work in order to quicken the emotional body. Whether the Higher Self portrays itself as a cloaked being, a suited figure, a triangle, a hill, a wave, or a lightbeam, all of its manifestations release in us the recognition of a broader aspect of that hologram that we call the "self."

Through these experiences our consciousness begins to expand. We free ourselves from our own human prejudices, our ideas of ownership and the prevailing of the ego. All restrictive thought forms that hinder the multidimensional being in its development fall gradually away. Learning this new language of dialogue with the Higher Self, we begin to change within. Our mental body changes itself as well as the emotional body. We experience—perhaps at first only during high octaves of consciousness in sessions—that positive attitudes and more pleasant, light-filled feelings and thoughts are possible! The old laws of the ego, defense and attack, begin to disintegrate and are replaced by all-encompassing principles belonging to a greater reality.

Subsequent sessions take one deeper into the reaches of our being, peeling away more and more restrictive layers. Whereas the first sessions usually focus on present-time experiences, these later sessions view other lifetimes, other cultures and societies, other dimensions. It isn't important if one doesn't believe in reincarnation. Regardless of how the pictures, situations and events arising in oneself are classified, the point is that they concern one's self. Our Higher Self will take us via the emotional body on profound adventures in other realities. We recognize these vignettes arising from somewhere within ourselves as meaningful and real. We automatically see the relationship these pictures have to our current everyday life. We recognize family members, friends, acquaintances. We see similar or identical situations to those with which we are currently involved. We recognize identical missions, themes and challenges.

Because these inner experiences are so relevant to our current lives, the voice of doubt within us is stilled, adding new truth to experiential living. We cease struggling on the level of linear understanding: "Is it really a past life?" This intellectual argument loses meaning for us because the quality of the energy experienced concerns who we are here and now.

Past-life vignettes arising in the sessions are always of the deepest relevance to the client's present-day life. Often current companions are clearly identified in other time-space situations, resolving misunderstanding in current life situations. The argument about the validity of reincarnation becomes unimportant as we miraculously get rid of seemingly insurmountable blocks we've carried for so long. The experience gained in sessions alters our aura and unconsciously influences our body language, speech and thought-forms, sending out messages into our environment. This in turn influences and transforms our daily life.

These inner experiences are in no way isolated from our daily life. They don't separate us from our children, family, the people we love, our religion, our bosses, our colleagues, our work—in short, from anything important to us.

We work intensively on the "themes" that have been elucidated by these vignettes. We gain an understanding of the principles that are affecting our lives. Often we discover that we are both victim and victimizer in traumatic events. We concern ourselves with these themes to finally release ourselves from them, to open up higher octaves that aren't dominated by our antagonistic survival-motivated emotional body. The second series of sessions is devoted to clearing our parents first, and then all those other close associations that set the tone of an emotional life.

When our emotional clearing has been sufficient to extend our attention to our life's purpose, we move on to the advanced "kundalini enlightenment" and "channeling" sessions. These are energetic sessions that quicken the vibratory frequencies of the chakras, amplifying the intake of prana, pure life energy. They increase our connection with divine energy so that we can begin to experience ourselves as the natural conduits of those energies into this plane. This serves to open up our life's path. Whereas before the themes tended to be focused on issues of revealing blocks, these themes begin to delve into scenarios of power and knowing that can enhance our global consciousness in this life. Memories of talents and gifts surface that seem to thrust us out into the world so that we can problem-solve and participate on the level of time manifestors.

The channeling sessions are the highest octave of the "windows to the sky" and are focused on creating new "grooves" or channels of thought cognition in the brain as it orchestrates our expanding consciousness. Making contact with new forms of interdimensional consciousness, we are able to further our own creative awareness.

It is our task in life to find out who we are, to become familiar with and use the power with which we were born. Why are we here? What do we do in this life? Were we only born to live, struggle and die?

We were born to open up the creativity within us. We have come here as part of a solution. Everyone has come here with a gift to give. Every single experience during this lifetime and all the other lives has forged us, refined us, so that we may burst forth and manifest who we really are. Here the emotional body can experience ecstasy, rapture and joy. In the channeling sessions we can direct our attention toward future possibilities of creative fulfillment, beyond the solutions to present and earlier problems.

Each time the emotional body experiences spiritual creative energy—even if this rapture is brief—positive effects endure in all areas of life. The emotional body orients itself toward repeating these experiences whenever possible. It becomes evident that we really can manifest beauty, artistic talent, scientific cognitions, elimination of difficulties, and loving relationships with people if we only allow the creative flow to enter. Once we've oriented ourselves in this direction, this creative flow automatically strengthens and carries us further. These are the octaves of miracles, of profound synchronicity that totally and so beautifully transforms our presence.

What happens when we return to our everyday lives after our sessions at the Light Institute? We are in the position to maintain contact with our higher selves. We recognize our higher self in new ways, learn to associate with it and to develop trust so that accessing our center becomes a natural way of life. These higher states of beingness are the birthright of every human being and can be attained without having to take on rigid lifestyles or perform rituals. At the Light Institute we help individuals find the center within themselves, wherein they can stabilize and strengthen their connection to their Higher Self. With the new insight and strength gained, we can move forward in our life and achieve whatever we wish.

We had a saying in the Peace Corps: "If you don't work yourself out of a job, you've failed." This meant that one should always help people to recognize their power to do something for themselves. The basic purpose of our community-development work in the Peace Corps was to help people help themselves. At the Light Institute we abide by this same principle: the divine presence of the Higher Self in our lives brings an ecstatic source of knowing, which is all we need to become whole beings. It is an illusion to think any external force, be it relationships, work or even guidance, can do it for us. The answer always lies within the self-created experiences of our multidimensional being.

ECHOES
FROM
OTHER
DIMENSIONS

4

Every individual has access to the Akashic Records, which contain, on an ethereal level, all experiences and points of development in the past, present and future for every soul. The more multifaceted our response to living is, the more tools we have at hand to shape and transform our human experience. It is simply impossible that these incredible resources could have been gathered by us in one time-constricted lifetime.

All of nature is a repetitive pulsation, repeating the pattern over and over again—continuously dipping back into the source to create the form. We are not an exception. It is divine consciousness that permeates all the bodies of form, and because we are a part of that consciousness, we are also a part of all form and matter that has ever been or will be.

Just as the echoes of our childhood prevail within the adult, so the echoes of all our multidimensional realities come to dance upon the stage of any particular lifetime. These echoes are the source material from which we mold the positionality of our prejudices, our passions, our predilections for lifestyles and even our physical bodies. How often it happens that the football fanatic, for example, passing through the dimensional veil, finds himself in that same Romanesque body—the gladiator, the Viking, the one who urgently must emerge the winner, for the cells remember the consequences to the loser. Because the cells do indeed remember and bear the interconnecting threads of all experience, we, all of us, can easily access those coalesced clumps of reality we call lifetimes.

Through the guiding wisdom of the Higher Self, and the astrally charged wellspring of the emotional body, these echoes are readily brought forth into consciousness in integral units. The mind-body's belief system is irrelevant. The truth is that one doesn't live just once. One isn't born to struggle through

life and disappear with death only once. Our customary frames
of reference simply cannot encompass the varied and seemingly
inexplicable imprints we carry in the emotional body that are
truly dictating our lives. We are ready, now, to discover the
whole of what is within, so that we can participate more
wholly in what is being shaped without. The wish, the longing,
the intention to discover the inner self suffices. Vital experiences about ourselves, our past, causes and possible solutions
to present problems, and opportunities for development in the
future introduce themselves as a matter of course.

We have all incarnated in every familiar race. As part of
the collective consciousness of man, not necessarily each individual, we have all lived in each key epoch. We can even
perceive these past adventures simultaneously with our present
ones. These earlier lives are stored within us and are part of
our collective memory pool.

The major epochs brought forth themes that were designed
to balance mankind and allow us to take our place as co-creators
within the universe. Atlantis was a major event in the myth
of separation. It not only produced male/female polarities, but
also genetic manipulation, and through its fascination with
technology amplified the separation between the spiritual and
the physical realities to the extent that man deluded himself
into thinking that his personal will was All. This was very
prevalent in the Middle Ages as well, when we became lost
in the tyranny of alchemy. The misuse of sexual power was
at its highest because the mastery of life force held within
drove us to attempt to penetrate its essence. The experimenters
literally disemboweled the body, in vain, to capture some
inexplicable force held within. It was the same force promised
in the Christ epoch, which was also lost to the symbology of
the cross. Instead of unconditional love and spiritual laws of

transmutation remaining as impressions upon humanity, something entirely different emerged with the symbolic cross. The
universal archetype of the cross was like a barrier for people
and became an imprint of self-sacrifice and martyrdom.

The cross portrays a symbol empowering the consciousness
to lock into or hold tight to matter. The cross fortified the
long-standing influence from Atlantis and Egypt. The self found
itself separated from God, as though its perfection were splin-
tered and worthless.

The originally intended message of love and perfection was
misunderstood and reversed. God and we were conceived as
very different from one another, and separated, rather than of
the same source.

Instead of creating light bodies, or "ascended" bodies,
which manifest themselves out of divine energy, we have adopted
the idea of self-purification through hell-fire and damnation.
We see the self as impure, as removed from the light. As an
external power, we fully accept God but deny the permeating
presence within us. With the exception of a few mystics (who
were often enough expelled from their religious communities),
we've impressed in ourselves martyrdom instead of love and
ascension as the original message.

By the time these negative concepts became second nature
within us, we were already too entrapped in the entanglement
between the victim and the victimizer, who measured out the
punishing price of divine worthiness. These roles were dramatized to the fullest during the Middle Ages. This was a
downward-spiraling path toward coarser material forms best
summarized with the word "Inquisition." Bodily crucifixion
and human sacrifice mocked the original message of Jesus worse
than any antichrist could have done. During the Inquisition
and in the name of Christ, people were brutally tortured in a

physical attempt to grasp or change something spiritual. Later, this intention was dramatized again in the alchemy labs of secret cults.

Currently we are in a reverse-rotational phase: in the upwardly spiraling movement of spiritual development. The rise of materialism during the Middle Ages provided for its own reversal toward spiritualism, in line with cosmic laws of movement of energy. Spiritual alienation had, with the exception of those few enlightened people and mystics, reached rock bottom. Ultimately, whispering formless spiritual energy began to break through rigid, encrusted shells of materialism. With this awakening, the consciousness was again ready, after the crumblings of alleged truths, to open itself to other realities.

The downfall of dogmatic barriers in our development of consciousness inspired visions in people. New ideas and plans are no longer forbidden or persecuted. Mankind imagined itself flying, and eventually did fly. We envisioned ourselves on other heavenly bodies and achieved it. This concerns much more than the intellectual mastery of material laws. The consciousness is opened up to the "impossible" and "unimaginable"! The spiritual awakening continues to spiral upward, unfolding new dimensions of consciousness, allowing new frontiers of manifestation.

Although we are still caught up in old-young concepts referring therefore to "victories" in outer space, or "control" in the air, the real revolution is the manifestation of the feminine power expressed by the intangible. As the first astronauts stood on the moon, they experienced God! They had broken through the gates of impossibility. The unity with creative divine force is felt through every body cell during the moment of breakthrough from the still impossible into a new reality.

Separation and difference vanish. We simultaneously relinquish personal power and enter the universal consciousness. We lose the demand for a separate identity and become one with creative power.

Regardless of whether one was victimizer or victim during these times, one will find the energy fields of these eras in oneself. As each era is clarified, one by one, during the sessions, the archetypal themes are released. The spiritual influences of ancient Egypt, Buddha or Christ were planted like seeds in the consciousness field of this planet, germinating in the balancing of yin and yang energies and spiritual clarifications still ongoing. Archetypal powers of these eras symbolize themselves in the clients' inner picture world so that they can see themselves as participants in the events and clear these residues that do not further growth.

These epochs are so important to our evolution because they concern contacting the divine power within us. We may see ourselves standing at the foot of the cross or experiencing the events of Golgotha. In this way we are deeply touched by epochal climaxes, wherein cosmic principles become tangible. This spiritual power was consciously suppressed in Atlantis, as the desire to dominate math took the fore. In Egyptian temples its use was secretive but less restricted. During the time of Christ, the attempt was to make these spirual powers accessible to humanity in general. Yet again mankind was swallowed up in the physical drama and unable to perceive the magnificent demonstration of cosmic law, which offered total transformation of this reality. The people who really understood Christ's message recognized that every living thing was a part of the divine force and that every person's birthright was to become one with God.

The archetypal epochs, for example Atlantis, ancient Egypt, Christ, Middle Ages and Modern Age, have fostered our development, each unique in the experiences offered.

We must come to grips with the problems of personal power, just as we have with mystical practices. We have tried to solve the dilemma of polarity between spirit and matter. These aspects of life are components in the repertoire of all individuals on Earth today. We are here to learn from each other—together. We have progressed far enough to comprehend the entirety of mankind. We have learned the lessons of the body, the instincts, the emotional reactions, intellectual research, religion and mysticism. Now we are called upon to integrate these lessons and take a decisive step toward reaching a higher octave, becoming masters operating with creative powers in this earthly dimension.

By using the past-life format to access these specific themes, we are able to reshape the magnetic astral connections that allow them to continue to play themselves out in this lifetime. Decrystalizing them as they have become encapsulated in this archetypal and obsolete form, we free the essence of the soul's energy to move again through us, so that the power and knowing available is reusable and even more perfected. As we release these epochal and dimensional memory packets within ourselves, we trigger their release in others around us whom we have attracted into this lifetime, through the very same energetic exchanges of those experiential echoes.

Let us peruse some examples of these dimensional themes. The following extracts from sessions are only mosaic pieces out of the whole pictures. Names and dates are changed or omitted to protect the privacy of our clients.

· · ·

Karin was working on feelings of being separate from God and not trusting in the perfection of life's experiences. Also there were underlying feelings of fear and anger toward humanity that were blocking the healing energy of love, which she felt was within her. Here is how the Higher Self explored this theme with her:

▶ "I'm a young girl, eight years old, and I've just come to a big temple on top of a mountain. It's an Essene temple and I've come here to be trained. I'm a wild and free spirit and prefer doing most of my learning in the woods, being with the plants and animals. We learn how to use herbs to heal, how to channel energy; we spend many hours in prayer and in meditation. At seventeen I am very radiant and channel light down through my hands, so that it feels like fire coming out. The healing begins with the physical body, which then begins to heal the spirit. Our community grows, and we talk and prepare for the god to come. We work on healing ourselves and strengthening ourselves for the times to come. I feel myself much older and I am in the presence of Jesus Christ. It is very wonderful for me, it feels like everything we've worked for has come about and he's there, radiant and beautiful. I feel surrounded by love. We women follow him and take care of his needs. It's this incredibly warm and powerful energy surrounding me, almost like a shock. It is like going home, it reawakens the memory of being with that source. It was a wonderful time, traveling, listening to him, watching him give to all these people. It feels like being a sister with him. The whole group is close. We are not treated differently because we are women. With him there is no fear or separation. My head hurts. It is the feeling that on the fringes of our group is judgment, denial, anger and confusion, coming from the priests and the authorities. The negative energies press in closer, and Jesus talks of the time when he will leave. I try to deny that, not wanting to believe it, yet fearing it. I try to trust in the perfection of God and try not to get scattered by the worry. My head is really aching. I am having a hard time understanding why the heaven on earth is not happening now. I don't feel anger and hatred for mankind, just a tremendous sadness and pain in the heart. A feeling of a tidal wave, of overwhelming masses of negative energies taking over. The whole crucifixion and death is like being engulfed and swept away. I feel panicked. I am near the cross, on the side watching. I'm in a state of shock. I can't believe this is happening. He is still up there, sending out light. I don't understand it. A feeling that we really blew it, that I'll never be able to experience that closeness again. That no matter what I say or do, it doesn't matter. I feel doubt, separation, hopelessness. I am no longer able to look. He hasn't died yet, but I go off by myself. The pain is unbearable. Years pass of aimlessness, of wandering, of not being present in the body. I feel separate, completely cut off from any nourishment or love. I do not feel I can trust God, or mankind, or myself. Death comes to me on a mountainside and at last I am able to reconnect with that ray of light."

This lifetime is a universal experience, in seeing how the emotional body imprints only the pain and the separation of the people, rather than focusing on the ascension and the great teachings and opportunities whose imprints would have transmuted the physical reality.

Margarite, a woman in her mid-forties, came to the Light Institute looking for some understanding about her relationships, particularly her relationships with men. She had never been married. Most of her experiences with men centered around having brief, intense affairs, which then ended abruptly, leaving her in a state of pain and loss. She had a tendency to greatly fantasize about the man she was with, to form illusions about who he was and what he would do for her, and not really see him as he was. She also chose men who were generally unavailable, such as married, already committed men, who lived out of town or were just afraid of being close.

Reliving several lifetimes with the central theme of relationship, we disclosed a fear of intimacy on Margarite's part. She now had imprinted in herself a fear of getting too close and a fear of loss, which was, of course, drawing these men to her in order to play out her scenario. One of her past-life sessions revealed a lifetime where she had a beautiful love relationship with a man who literally rescued her from being a servant, swept her off her feet, and carried her to his castle to live happily ever after. However, her handsome husband was soon killed in a battle and she was left with her enormous grief, sorrow and loss, which she then carried the rest of her life. [The lifetime was in A.D. 800]

▶ "We get to a forest and travel along a road that leads to a castle. It's high up on a hill. There are a lot of people who come and greet us. They take the horse. Oh, it's a beautiful place. He's showing me the whole castle. Everything is so huge! They are preparing a big meal. He dresses me so beautifully and he takes me to the table, and says, 'She's going to be my wife.' It's like a dream. I can hardly believe it! It's me, here. Me, here. I can overlook a very beautiful valley, forests and rivers. It's really wonderful, and I'm so happy with him. Then I see weapons. I hear noise. I dash down the stairs to see what it is. He tells me to stop there, says I'm in danger, I should stay there. He says he has to leave me for now, as they need him. I think that I'm pregnant. He comes up the stairs and holds my belly. I think he's got tears in his eyes. He hugs me and leaves. He never returns. He never, ever returns! He dies on the battle-field. He dies so young. We just started to live. So happy. I'm wearing black and mourning a lot. I wish him to be here. Everybody is very sad that he died."

She dies an old woman of stomach cancer, having lived only through the son she bore alone. Seeing the son is the only distraction to the grief in her life. She had suffered the loss of her husband for the remainder of that life, never allowing any pleasure to come to her. The message of the Higher Self given her at the end was, "I need not have mourned so long. I could have loved somebody else after my husband. I was free to do whatever I wanted to do. I was no slave anymore. I must take life as it comes and not hang on to something that has finished. I can be happy and fulfilled myself."

This lifetime enabled Margarite to see and release a long-standing pattern in her relationships: Fantasizing about a man who wasn't real and who lives only to rescue her from her problems and then—out of her fear of loss—keeping any man at a distance who could really love her in a nurturing, supportive way. She was able to release the pain of that loss she had been carrying in her stomach. She was able to see that she was creating this scenario over and over out of her need to see the love as being outside of her self, and her own lack of loving and nourishing herself. She came to experience that she had the capacity to feel love and happiness from within herself by connecting to her own divine Higher Self. This released the fear of intimacy and closeness, which will now allow her to draw in a long-lasting relationship that is real.

This is a classic example that demonstrates how most of our relationships have been fueled by a desire to fulfill our fantasies about what we need in order to be happy; and at the same time, our fear of that very intimacy and love we so long for. As we connect more and more deeply to that divine source of love within ourselves, we find that it contains everything we could ever need or want—all the love, support, gentleness, understanding, wisdom, security, strength and purpose. Then relationships can manifest at a new octave of deep sharing, deep intimacy, allowing a merging as the energy of love flows through the individuals in an ever increasing, expanding way.

Here is the abbreviated version of Mark's past-life scenario, which very profoundly changed his reality today.

"No matter how many times I went up in the helicopter, I was filled with a terror that would not be quieted by a rational mind telling me that I knew how to do it, as I had done it before many times. After Vietnam, I continued working as a medic in the emergency ward of our local hospital. I seem to have a driving desire to help sick people, and yet I get caught up in anxiety, a quiet panic about their getting well again. There is too little peace in my life."

Here is the Higher Self's explanation of that blockage.

▶ "From the time I was a young boy in the kingdom, it was recognized that I had the healing talents. I could speak with animals and plants and was accomplished in my art by the time I was sixteen, when I was taken to live in the palace of our ailing king. My attentions to him seemed to be effective, and he grew more contented and healthy by the day. One night, however, I was hastily summoned to his bedside as it appeared his heart had failed. I put my hand on his heart, but to my dismay, he died. Before I could even dress the body, an angry crowd appeared and hauled me away. I was led to the edge of a steep cliff, and in agonizing shock and horror, I was thrown off the cliff."

After this multi-incarnational (past life) experience, Mark was able to see why, on the one hand, he was so drawn to healing, and why, on the other, being a medic created a constant source of stress in his life. Even though reliving this heart-wrenching story was a deeply emotional experience for him, he was able to see humor in his fear of the helicopter and in his personal attachment to a patient's recovery. What followed when he returned home to work bespeaks the true purpose of this work. Here is a letter from Mark:

▶ "Back at the hospital I am feeling a powerful change in my capacity to trust myself in crisis situations. Last week a man was brought in with heart failure. The doctors tried all the usual methods to bring him around, but nothing worked. As I stood watching them, I began to feel a strange tingling in my arms, followed by a heat in my hands. The doctors left the room and I was left alone with the unconscious patient. Before I knew what I was doing, I stepped close to him and placed my hands over his heart. I felt almost as if I were in a trance. As I allowed my tingling hands to rest on his body, he suddenly opened his eyes and looked at me! In dazed amazement I realized that my body remembered those strange electrical sensations I had experienced as that boy healer in the life with the king. I am free again to use them, and I know that the cliff and the helicopter will never happen to me again!"

Alicia was working with the fear of becoming powerful and successful. In releasing the following lifetime, she was able to let go of her guilt and embrace her power.

▶ "I am wearing a beautiful gown, a royal robe. I am a tall, stately, powerful ruler of a matriarchy. I can see my throne; I'm standing by it, surveying what seems to be a palace. I'm extremely beautiful. There are crystals, large crystals, some with single points almost as tall as I am. I approach one and place my hands on its point. I take its energy into my body. I fill my body with its energy and power. I place my hands on my face, bringing its energy in and down through my shoulders. I use my hands to rejuvenate my body with crystalline energy. I then lie down on top of one and straddle it, throwing my legs and arms around it. My whole body begins to tingle. I tell myself that I can have anything, that life and God are my servants. I am incredibly arrogant; all that is in my life is put there to serve me. The matriarchy—I have placed myself above

God and see others as the slaves of my whims. I have harems of men, beautiful men with chiseled bodies. Their job in life is to make their bodies beautiful for me. They are my servants! I see all these people making beautiful gowns and jewels. There are soft cushions for me to lie on to experience the exquisite softness of my form. My kingdom is very wealthy but there are many who are hungry and who barely make enough to live on. I make them bring their food to me; only I am important. They are here to serve me. The people here bow and pray to me for their harvest, and for everything that they wish for in their lives. If people do not see me as a goddess I have them tortured or killed. I am completely overcome by my arrogance and sense of self. I am incapable of seeing what is before me. I am so wrapped up in my pleasure, sensuality and pride. I have two or three men make love to me at one time. I drink their fluids into my body. I take their energy and keep it. I do not give it back to them. I only receive. I delight in watching them orgasm and orgasm and taking from them their power and their youth. I will not allow them to be with anyone but me. I have them play with each other when I get tired. I use their energy to enliven my body. I do not orgasm; I just take their energy into my body. I do not give back any energy. My body stays young through the energy of the crystals and the men. My heart is cold, like the heart of a crystal! I die on a soft couch with my men standing around me, praising my beauty, but in their hearts they are so happy that I'm leaving. The whole kingdom inwardly rejoices to be rid of me. There is no peace on the other side for me. My heart opens and I must look down and see what I've done. I see the people I've murdered, hungry children, the abuse of power, the lie of arrogance, ego and body. Beauty is not in the body. Beauty is an open-hearted compassion. I feel the misery that I have created. I do not trust myself. I do not trust that I can wisely handly power and beauty. I am frightened by power and so afraid that I'll do this again.

"In my next life I am choosing a simple life, devoid of power or knowledge, to hide from myself. I can see myself in the Middle Ages, dressed in rags, deformed, missing a hand and covered with sores. I am a leper, despised by people. I am in the mountains. I am in much pain and it aches even to walk. I am so alone. I have much time to think, to be alone with my pain, my deformity, my ugliness and my misery. I have come to love simple things like wildflowers, little brooks and the handouts of the kind shepherdess. I look at her from afar. I do not want to show my face for fear that she may be frightened. It makes my heart glad to see her. I am thankful of the bread, cheese and goat's milk that she brings. I give thanks to God.

"It is such a blessing to let go of the body, to be done with this life. My heart is so much lighter now. My ego has judged me so severely for my evil. God does not judge me."

This is a man's advanced session in which the Higher Self gave a profound experience of oneness.

▶ "I feel a very pure lavender color—soft and heavenly. I feel/see a very soft, unearthly pink color combined with lavender, like the soft pinks seen in a white opal.

These colors and energies enter my body underneath the chin in the throat region. They have filled my head and chest.

"Since the colors became visible I have seen/felt the most beautiful picture of a being in human form, unlike anything I've ever seen before. After all is said, I still can't find the right words to describe this person. I look straight through him/her and the main thing apparent is beauty—godliness. I'm not sure if it's a man or a woman, but it's very pure and clean. We're communicating with each other. She says that this is the form I had in a past life. It is pure energy flooded with light. It's a man, but it's so beautiful I mistook it for a woman. The sex makes no difference. It told me that this is myself in the present. It is hard to believe that, because this being is so beautiful and pure. Pure energy. Consciousness and God.

"It tells me that this is the form I will assume in this life on this current level. She/he radiates wonderful light, the most beautiful light imaginable. Glistening and soft, intense and pure. It is a picture of God.

"I am told from inside myself that everything that I do and touch shall occur with and through this light. Doubt and failure do not exist. I shall go through my activities with this knowledge. I shall continue what I've been doing up to now, but with another energy. There is no separation between earthly activities and the ethereal. Up to now there was always a separation between daily activities and ethereal realities—this schism is now gone. Daily life will be elevated to ethereal levels. Tears fill my eyes—too much beauty. Breathtaking— beyond word. No words can describe this experience.

"My physical body and this light merge. It is astounding how everything perfectly harmonizes down to the last cell. My physical body takes on another form.

"The point where we merge lies in the chakra underneath my navel.

"I have a job, routine, but my main responsibility is to radiate this light and expand."

[Where did it come from?]

"It came out of pure God consciousness. A direct light-stream out of the well of God. It came without lifetimes of karma. It is full of all wonderful colors, none dominating the other. This light body was always there, but buried under layers of karma. Experiencing my light body merging with my physical body melts away the karmic mask.

"It is significant that in this past life I'm perceiving myself as creator of whatever I dealt with. I could manifest everything I wanted with the help of my light body: space, food, buildings, total manifestations of thought. I need not select any single activity, but instead create everything through thought processes. While I project my aura, I can manifest everything: harmony, protection, simply everything. My presence penetrates all existence. Every cell, every molecule of my body is lit up with light."

An especially exciting aspect of our work at the Light Institute is helping mediums, healers and psychics clear their emotional bodies. The more clarity such people gain about themselves, opening new dimensions of consciousness, the more they can help others. The further their vision goes the fewer are their prejudices and positionalities. During the last few years we have worked especially intently in assisting those bodiless, disincarnated spirits that work with and through the mediums and healers. We establish contact with these entities and let them tell their stories. This helps them become more aware of themselves and clarify their emotional attachments to this plane. They are then in a better position to help others from a higher spiritual standpoint. Without newly incarnating they can free themselves from the limitations of their life experiences. Many times there is a symbiotic relationship between the channel and the entity who may have "unfinished karma" on this earth plane, which can be balanced by working through the channel who in turn is able to utilize his or her special talents more fully under the protection of a disincarnate being. They can assume higher viewpoints. It is fascinating to help not only embodied humans, but those disembodied spirits as well.

A very well known American spiritual medium came to the Light Institute to work on her emotional body. During one of the sessions some of her spirit guides announced themselves. I asked who among them would be interested in clarifying their own problems in order to be more effective, or to actually release themselves from their karmic duties on the earthly level. One of them responded, introducing itself as "Mother Venus." She carries an energy similar to that of Mother Mary on earth; a vibration of motherly, caring love and harmony. I asked Mother Venus why she turned toward the people on Earth. She said that she experienced pain from the suffering of the earthlings, be it starvation, catastrophes, earthquakes or war—all leading many people to horrible deaths. She was compelled to help.

I advised this Mother Venus to glance at the Akashic Records to find out why so many people died under such conditions. I cannot describe in detail all the breathtaking realizations during this session. Mother Venus was able to view simultaneously the records of many thousands of people. She determined that every single soul dying with a horrible suffering death had chosen this itself to gain a certain development of consciousness. She recognized the source of the decisions of such souls: to gather experience through certain conditions in life and thereby to grow. (So that no misunderstanding results, this insight shall not be interpreted to mean we need not relieve emergencies or help suffering people. To the contrary, the more loving we are to others, the more love can work as a life principle for our own experience and for all humanity.) The choice to suffer is always for the purpose of growth. The fact that those who suffer do it within our environment, our life, our movie, calls upon us to respond for our own growth. *We* are not disconnected—to the contrary, they may be suffering to help *us* grow. There is a creative purpose behind all occurrences. As Mother Venus realized this, she was able to release herself from the emotional identification with the suffering of those people; the whole room filled with light and energy. This spirit from another planet then chose to assist the passage of deceased souls, rather than to let herself become numbed from the pain of their death circumstances.

The woman acting as a channel for this spirit fell into the memory of a previous life at the moment she spoke Mother Venus's explanation. She saw herself on a stretching table in a torture chamber. She was tortured in unimaginable ways because she had spoken to the farmers in her town about spiritual things. She had said that everyone was a child of God, that everyone could establish contact with God without turning to the priests. She spoke about ascension. The Roman Catholic dictators grabbed her and used torture to persuade her to renounce those views. The farmers were helpless to come to her rescue. She refused to take back the purported heretical declaration that everyone could cultivate a direct connection with God, so she was tortured, quartered and died miserably.

With this inner revelation, the woman realized suddenly why she functioned as a medium in this life. Her body remembered that it had paid for her having spoken the truth and simply did not wish to experience being put to death again. It appeared to be safer to pass on higher truths that, although part of her own multidimensional beingness, she could relay from "other" sources, whether those were disembodied spirits or her own intuitive knowing. Such spirits are, however, not always separate beings from us, but frequently an aspect of our own consciousness. The theme of undergoing terrible punishment for revealing knowing is classic of channels and mediums. Once a soul has gleaned divine wisdom, however, it cannot retreat from those lofty octaves and thus finds its bodily vehicle over and over again in the position of the messenger, often with the same terrifying results. Man has always found things unexplained to be threatening. Fear is the precursor of killing. She reports that the quality of her channeling now is from a much higher, clearer octave.

People often have very positive experiences going to mediums or channels, who relay helpful information or guidance from the spirits contacted. Sometimes this helps them recover from the pain of losing a loved one to Death or from the heartbreak of a crumbling relationship. They may be assisted in business decisions. Prophets were relied upon throughout the history of mankind. It was wonderful to benefit from the talents and offerings of those beings who had developed these capabilities.

It is astonishing when someone else knows something like this about us. In our surprise, we believe we have encountered someone who knows more about ourselves than we do. We presume that because they know what we have experienced, in some magical way they will make better choices about our life than we will.

Spirits are nothing other than disembodied personalities. The prejudices, judgments, viewpoints and ideas created by their karma during their incarnations are still embedded within them. If we ask them for advice, they will reply based on their own limited experience from their previous incarnations. Some of the spirits haven't embodied on the Earth in possibly hundreds of years. Their level of consciousness and cognitive capabilities is thereby subject to the realities they exposed themselves to then. We are too ready to forsake our self-responsibility, avoiding our own cognitive ability, intuition and volition under such impressionable conditions, hoping to somehow avoid our karma through the abdication of our power to someone else. In short, the down side of channeling lies not with the channels themselves, but in our addiction to externalizing and our determination not to be the one responsible, that is, guilty. We always profess to be the innocent bystander—even when the scene is about ourselves! At certain times a channel can be thoroughly helpful. Some of them play a very positive role on this earth, individually or collectively, especially at this moment when mankind so desperately needs a global perspective of his role on this earth. This is much different from titillating our personal curiosity.

We are all surrounded by spiritual guides. The angel stories about heavenly beings and cosmic figures assisting us are true. There are a multitude of spiritual guides that have decided to float through this life with us. We can actually establish contact with them and sometimes they are useful.

The path of my life was completely altered when a channel guide first showed me how to trigger consciousness through the threshold of multidimensionality. Soon after I began with acupuncture, a spiritual guide named Ling Shu announced himself. He began to tell me about the existence and exact layout of esoteric acupuncture points. Often it happened that he spoke about points that I didn't recognize as "official." I had to first try out his information to prove its validity. It was an act of trust in Ling Shu to place needles in esoteric points he had indicated. After crossing the threshold of trust, I could research and prove the validity of his assertions. His substantiation of the energetic location of many points, which hitherto had been unknown to me and others, left me spellbound. My Higher Self played a hand in this confirmation by leading me to copies of one of the most ancient acupuncture texts, which was entitled Ling Shu. I could not miss the synchronicity or the humor!

Ling Shu is no longer so frequently with me, though his presence in the healing room can sometimes prove not only enlightening but embarrassing. I can almost feel his cosmic breath of attention over my shoulder. Many times his guidance has taken the form of a strong insistence that I place needles here or there. Many times, either he felt me a slow student or decided to add a final touch of his own; he would place needles himself while I sat by scrambling for explanations to my clients. Once a client lay on the table, and while she voiced her inner experiences, I sat at the head of the massage table taking notes. She suddenly asked, "Did you just now put a needle in my foot?" I assured her that I had been taking notes and saw no needle in her foot. Then I discovered that Ling Shu had placed a "needle" from an astral level to support and strengthen my work. Even from these subtle levels the consciousness of the client is opened and intensively activated.

Spirit guides can be wonderfully effective and helpful with their unusual talents, whether in acupuncture, music, art, science, earthly or spiritual knowledge. However, their knowledge, help and expertise always remain limited to the experience and consciousness level of their own incarnation. If one wonders, "Which path in life shall I pursue?" or, "What is good or bad for me?" there is only one source to turn to: one's own Higher Self. It is best for our spiritual growth to find direct contact with our own Higher Self, building, refining and intensifying this connection.

If one daily establishes contact with the Higher Self and lets oneself into this energy vibration, either through meditation or simply through short pauses while washing dishes, driving or before important decisions, this would be most beneficial. Everyone can do it! If we pause shortly, breathe deeply and simply ask our Higher Self to take form or to be energetically present, this suffices for a beginning. It is finally possible to take the first decisive step in this conscious awareness of the Higher Self. It is the first step into the tunnel that will finally pull us through a vortex of darkness into the light. People coming to the Light Institute are from all walks of life. They perceive or recognize that there is more to life than success on just physical or social levels. They know that the image they present to the world doesn't fulfill them in the end and that simply diverting the restlessness in themselves is not enough. They embark on a search without knowing what they are searching for or why. The seed of their search was never planted by anything outside them——it comes from the level of their own soul. This seed is determined by destiny to sprout forth in this lifetime, and therefore they set themselves on the path. It is the path to enlightenment, illuminated by the Higher Self.

As these people perceive other dimensions, past lives, reflections of their encrusted self-image, their letting-go becomes easier. They dig deeper and discover more about themselves. They begin to recognize the sources of many impediments and develop the readiness and capability to disentangle these causes. For example, often people come here with acute or chronic illnesses. If they encounter the source of their illness while experiencing other dimensions, their ailments and symptoms usually depart on their own. The rug is pulled from under the physical or emotional manifestations of conflicts and difficulties, since the source is now evident.

It may be that some people always seem to find partners who never love them, for example. As soon as clients recognize their participation in this through conduct in earlier lives, experiencing it on different levels of their multidimensionality, they can let go of this energy vibration during the session. They can release those hardened habits that have deposited themselves at the cellular level and replace them with a flow of healing energy that automatically generates love. When we discover how energies work within us, we can transform dull, unconscious or negative patterns into higher, lighter frequencies, which absolutely change who we are *now*. Multidimensional reality amplifies our choices and responses to ourselves and the world outside us.

TIME
IS AN
ILLUSION

5

Time is a sheerly unexplainable phenomenon, a supposedly eternal mystery. In our sessions at the Light Institute we experience the quality of time in completely new ways. The limits of our time concepts become apparent and we release ourselves from them. Present time becomes continually expanded. Time itself seems a qualitative disruptive aspect of energy—like a ripple, an arc whose motion is contiguous to itself as each incident of experience creates the waves that move it further and further out in the cosmic sea.

The more we discover about ourselves, the more we experience that our reality and the various ways of perceiving it know no time continuum, but speak more to the arc of the hologram that creates the integrative potential of multidimensionality. There is no indisputable chronological sequence of consciousness, or "conscious beingness." When we are in touch with our own essence, we break through the time barrier because this essence of our lives exceeds time and space concepts.

In our sessions working with the emotional body, whether we relive our childhoods, impressions from past lives, or find ourselves in other dimensions, the energy is perceived and effective in the here and now. At the onset we realize that past and present are not separated. We discontinue using these purely linear concepts as a means of experiencing our reality. Without a doubt, when a past life appears in a session, the energy from that life applies to us today! This awareness-expansion process can be compared to a stone falling into water, releasing concentric wave circles around itself. Our consciousness and perceptions expand to experience amplified dimensions and scenarios, and by practicing perception and assimilation of these realities, time loses its influence on us.

We are no longer compelled to struggle with something

yet to be, not in the here and now, because we are fully involved with the present. In touch with our inner essence, our life force, we realize that "now" continues like a ripple in the water of life. Once we have experientially accepted this, our consciousness is transformed.

As we come to realize the various roles we play in other bodies and other dimensions, our consciousness broadens. We aren't restricted to one dimension/one form/one time. Through my research on the emotional body, I have learned firsthand that time is an illusion. The emotional body is on the astral level, in which neither time nor space is recognized. For example, a childhood trauma is stored in the emotional body. Upon encountering similar or identical situations, even as a much older person, the emotions of and reactions to this trauma are called forth and dramatized anew. We have already cited the example of adults reacting toward their own parents as they did as small children—a clue that their emotional body unconsciously determines their behavior. The emotional body, existing timelessly, isn't aware that with adulthood such subordination under possibly oppressing authority is unnecessary. How one perceives himself is a matter of consciousness: eternally thinking oneself the child, suffering under external authority, or as a whole being—free.

We guide our clients through these earlier experiences, helping them release themselves from the coercive control of the impressions from these experiences. They are released, and discover their own inner and outer independence. They cease seeing themselves as children or as victims, but rather as complete people, entirely conscious. In that moment wherein they experience themselves as creative essence, formulating their own destiny, they prevail over time.

We need not die—we need not age! We already know

that there is no physiological necessity or reason for aging. Our consciousness is at the point of conceiving, but not yet manifesting, "Let the old cells die, let new cells be born, let the body rejuvenate itself." Science is looking for some unknown clock that will explain why the seemingly inexhaustible DNA system breaks down. The answer lies not in the body but in the divine power of choice to incarnate for whatever purpose and in the gift of teaching that death itself brings. When we explore these spiritual principles, we will have these answers and usher in a new age of consciousness—life as transcendence!

We can consciously raise the vibration of the body so that a "body of light" is created, steeling the cellular body. We can consciously renew the body, not out of fear of death or out of desire to enforce the will upon the body but because we can choose to orchestrate the creative art of bodily life and death in concert with the soul's plan. We can maintain the body in a timeless condition until we decide to "change vehicles." Knowledge about these things has always existed, and it won't be long before we will be able to incorporate it into our lives. All the great spiritual traditions speak of this timeless body phenomenon. Some examples are commonly known: Yogis who survive many weeks and months buried underground without food consumption, the "ageless" avatars who have timeless bodies, and the countless stories of people who have death experiences and return to their bodies to finish some spiritual task. Notice that the practitioners of such miraculous body art are all living on spiritual octaves.

If we have the courage to float through all these dimensions, collecting information, recognizing again and again new aspects of reality that exist outside linearity, time sets no limits. When one releases the kaleidoscope of experience, the "memory" of the cell "explodes." The self (the soul) moves according to cosmic law, always higher and nearer the wellspring of creation. Life moves itself upward, always toward the light. Instead of searching for a dull reflection of reality, we experience these states of consciousness directly, fundamentally transforming us.

This transformation emanates from the molecular structure of the cells. It accompanies the freeflowing pulsation of experience gained when the body is released from the restrictive, reactive patterns that were "engraved" in the cells. The borders of our consciousness are set by the emotional body, which retains, carries on and intensifies all reactions. When the body transmutes the anger, prejudice and other stored impediments in the cell structures, it becomes like light itself, awakening us automatically onto higher octaves. It is amazing to witness the visible transformation that occurs when we no longer wear the face of fear or anger. Since the body is no longer screaming for attention, it has no use for illness. A peaceful face is a timeless face! We experience happiness without struggling. This breakthrough is achieved in very natural, sometimes unnoticeable ways. We suddenly realize that a year ago we had to fight for career success, whereas it is achieved more easily now that we are able to recognize what we desire from life. Or we may observe that before we had only difficulties and grief in relationships, whereas now we are able to really receive love. As we consciously contact the timeless Higher Self, the cellularly stored fixations of the emotional body's feelings and reactions become nullified and dissolved.

What is begun in the sessions continues thereafter in a natural way. The physical body, mental body, emotional body and spiritual body begin to vibrate in harmony with one another. This penetrates from the finest subatomic particle through

physiological processes, feelings, thought and cognitive processes, astral and intuitive levels, up to the highest octaves of
being. The bodies begin to merge. The life force of these bodies,
the "shakti," pulsates synchronistically. As we come into direct
contact with the life energy in our cells, in our emotional body
and in our spiritual beingness, merging these energies, then
we too become revitalized. Our consciousness will rise to a
radiant level of joyfulness, which doesn't continually pass judgment or criticize or always stand guard. We experience what
it means to simply live and be alive. We needn't be caught in
the net of time because of the repetitive inertia of an emotional
body cut off from its divine Higher Self.

It is astonishing to recognize that the disassociated emotional body repeats itself at its lowest octave. Even though we
all have had, within our hologram, at least some experience
of merging bliss, we are unable to know freedom and take the
mantle of creative responsibility into our lives because we only
remember the experiences of denial and confusion with which
we so easily identify ourselves. Imprinted thus with negativity,
the evolution of consciousness falters—time stops. If we do
not remember or access imprints of ourselves as manifestors,
as powerful designers of our own fate, we do not have the
avenue of choice. Freedom is a mystery to us. Thus we repeatedly reenact the limitation with which we identify, even
though at a deeper level within us all resides the reality of
ourselves as powerful, knowing God-beings.

The passage of time is not real to the play of our emotional
body and has no influence on our inner imprints. It is an
illusion that we have changed. We may seem many years older,
but we still live within the prison of our choices, which stop
the clock—whether that be at the transition from childhood
to adulthood or at experiences of other lifetimes—and we

simply adapt for reiteration in this time frame. To evolve, to be free, we must disconnect those imprints which paralyze and consume us, leaving no energy to extend our choices to new growth-producing experiences.

I will pose my own poignant example of how old imprints can destroy our freedom of choice. Such a challenging, unavoidable theme in life originates through an accumulation of similar choices from different lifetimes. These "engraved" experiences effectually dictate our eventual real-life conditions and realities. The period of life I'll briefly describe was determined by a certain condition of consciousness concerned with the freedom of personal decision and self-knowledge. This created a very sustained determination of my way of life and beingness in this life.

It concerned a lifetime in Egypt. I was a young boy, destined to become the next Pharaoh. It was a life in which I, as in this lifetime, was born open. At that time I remembered knowledge and experiences accumulated in past lives. I knew that I had already gone through certain initiation processes that were prerequisites for admittance into priesthood. The priestly functions mastered internal realities as well as guidance and control of external realities. This was achieved through the application of special talents or powers.

Even as a young Egyptian child, I recalled these previously acquired talents and powers. These well-explored abilities were very useful to me. I could speak with animals—a wonderful, loving experience creating an inner balance. It naturally exists outside the limits of linear, restricted understanding. I could telepathically beckon birds to fly to me. I was capable of grasping the vitality expressing itself in these birds. We could have interchanges with each other. In certain ways we could become one. Communicating with a bird cannot be compared with ordinary audible speech, but is more like an intuitive, conscious, mutual understanding, a total harmony between us. I could also establish contact with plants in these ways as well. I was able to hold them in my hand and describe their characteristics and qualities. I'd know if a certain plant could be used medicinally or as a food. I'd know if it loved lots of sunlight and what other plants would grow harmoniously near it.

One of my earliest memories from that life concerned a certain garden near the Pharaoh's palace. On the east side of this garden were wonderful sandstone walls. My teacher, who had been with me since I was three years old, drew symbols on them and I could identify their meanings from my memories of studies during past lives. These hieroglyphics told the history of the different plants in this garden. Sometimes I'd draw information about the energy of the plants and their usage on the wall myself, but mainly my teacher did it. These walls glistened from the sun's reflection on the high silicon content of the sandstone. I felt very vibrant then; it was a time of enjoying the beauty of life, very much like a game. My teacher and I had a sort of cat-and-mouse game between us. He would point to a hieroglyph, one not yet discussed in our lessons, and I'd sink deep into my memory and joyfully ascertain that I could activate my knowledge learned in past lives.

This dear teacher was also my educator during adolescence. He taught me intellectually and spiritually. He helped me realize who I was and who I would become in that life. As the time for my multiphased initiation drew near, wherein I would prove my knowledge and myself, I did not feel the anxiety a student anticipating exams normally felt. Despite my knowing that I would either pass the initiation successfully and be destined as Pharaoh, or die, I didn't worry about failing because I'd never failed in that life. I was loved and successful. I spoke many languages. I was truly a fascinating child. Just experiencing this Egyptian child within myself healed a lot of the pain I had felt in my present childhood in which these talents were not appreciated and left me feeling odd and unacceptable.

I entered the initiation with full optimism and saw its completion and my part therein as simply a ritual. I was fully confident about the components and steps in the ritual and was sure I would master each step through to the finish, exiting as Pharaoh. I remembered the attunement and preparation in previous lives for this life as future Pharaoh. While proceeding along the path toward my initiation ritual, lapis lazuli, turquoise and gold dust decorated my eyelids, each carrying the energy of three different octaves. (After the session, I had a hard time opening my eyes and felt that dust in my eyes for three days.)

After careful preparation for the initiation I passed through many rooms by boat to an inner chamber. The testing began with very rough trials of courage, physical demands and tasks, and proofs of will power. Although it required the highest concentration to endure, I was able to maintain it through to the last part of the initiation. I was already bursting with joy because I knew that I'd arrived at the final step. I knew I had completed the ritual flawlessly, passing with flying colors.

Before I could step out into the light of day—as Pharaoh— I had one last task to fulfill. I had to make a decision. I saw before me a picture of a forked path. I had to choose one of the paths. With great certainty and joy I knew that the left path was the correct one. It was easy for me to identify from the inner voice of my intuition. Just as I wanted to voice my choice, the face of my beloved teacher appeared before me. His face was very instructive and worried, commanding my attention. The expression in his eyes suggested that the correct path was the right fork. This beloved teacher was the source of my self-realization as a powerful being. He was loving, with eyes full of empathy and affection. I saw myself as a reflection of him, admiring and trusting him wholeheartedly. I thought immediately that for some reason I had erred, and so I altered my choice to the right-sided path. In that moment a flash of light struck me dead. This event had such long-lasting effects on me that my spiritual body was in a state of confusion for a long time.

I had gone against my own knowingness! Out of love, respect, and because of the reflection in his eyes in which I saw myself, I renounced my own choice. I was intentionally "betrayed" both as a necessity of circumstance and to teach me a very important lesson. My teacher knew that in that lifetime I was bent on bringing through certain spiritual knowledge, for which the people were not yet ready. I was therefore removed. As Pharaoh, I would have made such knowledge accessible to the people, and my teacher, knowing this, moved me aside. It was an unforgettable lesson—that each being, regardless of the circumstances, must above all follow his own knowingness! Our own understanding, our own inner voice is the strength that forges our path in life. When we follow this inner guidance and consciously decide according to our own inner voice, we can never take the wrong path. We should never abandon our own judgment even if a path appears very misleading or dangerous. The path chosen by our own judgment will always be that which brings us the highest evolvement for the soul. Is this not in fact the reason for our incarnation? It is why we choose bodies and occupy ourselves within this limited dimension.

This lesson is significant, not just for me in this life but for all people living now in this time on planet Earth. I am reminded of my early childhood in this life. I was often shoved aside by friends, classmates or family members when they were disturbed by my obstinance in fulfilling something they considered unimportant or uninteresting. By reliving that Egyptian life, I understood why it was so important to me in my present lifetime, especially as a child, to stick to my own choices, even though I was perceived as obnoxiously willful. I often paid dearly by being ostracized by friends and family.

To see the parallels between that magnificent Egyptian boy and my own devalued childself completely reshaped my sense of self worth. In both lifetimes I was only concerned with being loved by those I loved. To remain steadfast by my own choices, even if it meant rejection, demanded that I must develop the courage to stand alone. If one really abides by his own decisions and follows his own judgment, one can never become a victim of circumstances. One's own role in life, one's own reality, takes on another quality.

OUR CHOICE:
VICTIM
OR CREATOR

6

"Who am 'I' in this life?" This question, the problem of personal power, surfaces in almost all sessions at the Light Institute and represents a challenge faced by all people living at this time on the planet. Our vibrating frequency is higher and faster than ever before. We have established rudimentary contact with our spiritual bodies, enough to search after the purpose of our lives. We have passed through millennia in various life forms, some very oppressive, accepting that we weren't free to develop according to our own free will. We believed that we were born to struggle through life and finally die—with no alternative, no free choice.

In the meanwhile our consciousness has broadened and opened up, so much that we now inquire about the reason and purpose of our being here. We needn't struggle with bare life necessities now. We may be financially comfortable, successful, and have a loving family. Despite that, we begin to think about why we're here, what our task in life really is. These searchings are a main concern of people coming to us. What shall they really begin in their lives? Which experiences would be the most rewarding? They are concerned about finding out who they really are and how they can manifest this in their lives. They immediately run into barriers or impediments in their search and in their own manifestations.

Some have problems in personal relations, with parents, children, superiors, teachers, priests or spouses. We form a dependence on outside sources of power, an alliance that, no matter how bitter, we would not choose to break—lest we be faced with the truth from which we are hiding, our covert mistrust of our own power.

The major obstacle to our growth is inertia. The emotional body permeates all our body consciousness with a warning to hold on, to maintain our crystallized reality at all cost. As it becomes habitual in interpreting new data it closes down most of the avenues of motion—namely, choice. Time stands still—revisits itself. Deprived of new possibilities, the body simply switches to "automatic pilot" and an inert numbness spreads throughout the being.

Numbness creates a passivity toward life. Energetically there is a lack of vitality, so that little energy radiates outward to embrace life. From this weakened position there is no resource from which to create. This situation predisposes us to the stance of the victim. It is literally a posture; a flinching that mirrors itself in all the subtle bodies. Its consciousness limits itself to the fight-or-flight syndrome—fear and anxiety. The victim never lives in the universal now, but always in dread of the future and the shadow of the past. Lacking the energy that provides the shell of impermeability and immunity to external negativity, the victim is the designer of his own self-held prophecies, whether those are the unspeakable fears of physical harm, or psychic self-destruction and failure or depressing doomsday attitudes toward the outside world.

The inertia leading to victimization is not an energetic way we are taught; it is a primordial response, echoing out for multidimensional experience.

People who choose the victim role very often have a theme of self-judgment relating to misuse of power, and self-hatred originating from some unconscious "unspeakable." It is quite a tribute to the law of magnetics that we are able to get some other being to perpetrate those "unspeakables" upon us and thus maintain that tenuous thread of innocence, simply because the emotion is coming *to* us—rather than *from* us. Viewed from this perspective, we can see the telltale link between the victim and victimizer.

The game of denying power usually begins by choosing

parents who will strive to keep us powerless in compliance with their own experiences. These stifling relationships must be understood as a reflection, coming back to us as externalized power and serving as a prototype, a form of influence or control that overwhelms us. The control lies in that authority being able to define who we are, how we should see ourselves, and how we should develop. This is, of course, readily observable in parent-child relationships. Parents impress concepts and judgments upon their children, and children make decisions based upon these impressions. The cosmic joke therein is that children have chosen their parents, using them like pawns on a chessboard. In this game the children choose the parents to reflect the themes and issues to which the child is most resistant or reactive. The purpose of this juxtaposition is to balance cause and effect on a karmic level so that we learn to take responsibility for ourselves.

A child may have a dominating parent, constantly critical or even physically aggressive. Or the child may never receive encouragement, but invalidation instead, i.e., "You are too dumb, you won't make it, you won't achieve anything." From the child's spiritual level this was of his own free choosing, designed to give him a necessary push toward the best possible development.

The soul envisions the abilities and knowledge needed to learn in this incarnation. Through karmic bonding and mutual connection between beings that child will attract those souls with whom he's been together before on deep levels. Roles are often interchanged, as the individuals take turns being father, mother and child. The souls are in mutual agreement to incarnate in synchrony, helping each other gain the experience appropriate for their development.

The child is never a victim of the parents. The parents and

child chose each other. The parents are principal players in the child's chosen life plan, which he himself continues to direct. Fascinatingly enough, in biographies of famous people, regardless of time or culture, those who achieved eminence had severe childhoods.

What makes us able to leave our mark on the world? What makes one into a Buddha, a Jesus or an Einstein? Adversity often helps forge what we are or will become because it triggers the energetics of power we already have! If we experience restrictions and limitations because of negative power, we are challenged to wake up and focus our attention on this dimension. Our awareness of and connection with the Higher Self provides us with a magnificent new tool for perceiving our life's lessons from the vantage point of the hologram, and thus making choices that integrate our experiences on the level of creativity. (In this new age, the rhythm of life has strongly accelerated, and the life forces pulsate stronger and faster; we will establish other methods of teaching ourselves.) Within the framework of the cosmic game and in synchrony with the development of life, people emerge from hard-learned lessons with special talents and capabilities, and with their own consciousnesses raised. Through the process of these trials, their souls are forged in such a way that they radiate a special energy in the world.

Out of the curled-up fetal position of the "victim," who asserts powerlessness, is formed a certain power, the tyranny of the weak. This power forces them to become aware of a specific quality of their beingness and to manifest it, radiating out into the world. Whether they radiate as artists, scientists, through the love in their hearts or something else, they break away from conventional invisible lives.

In our culture we have developed a social system in which

the "masses" work like complacent sheep. In this system we encourage a yearning to be accepted and loved, and the desire to fit in with others as a means of exercising power and control. Power and control mechanisms have been the major incarnational challenges on this planet. Every religion, government, every outside authority is eager to control people by governing their behavior. Up until now we haven't had any developmental model that is not based on either perpetrating power or acquiescing to power. In other words, we live in a system that calls for victims and victimizers.

At the Light Institute if people say to us that they had tragic childhoods or their parents always scolded them, making them rigid now or making them unable to love others or be happy or creative, we guide them back into past lives, into the source experiences in other dimensions of consciousness. We return with people like this to the source of the problem and discover why they allowed themselves to become victims in this life. All people acting in this life as victims have their own repertoire of roles played in past lives in which they were victimizers. Ultimately we are confronted with the vicious circle, the victim and victimizer are one!

The most profound lesson is that regardless of how they are victimized in this life, whether under brutal parents, relationship problems or whatever, the "victims" always ask, "Why am I with these people?" Amazingly, they learn that they wielded as victimizers the same force against those people under whom they now suffer. Even more shocking is their realization that the victimizer in this life is often a soul with whom they've traveled many lifetimes. The karmic processes I am describing here aren't punishments! Karma means merely balance and choice of experiences. Thereby a flow of energy originates, an ebbing and flowing of personal and cosmic tides that move the universe. Yet it needs other souls ready to enter this dimension as role players to balance the interchange of experiences. Many other souls will refuse to go along, having perhaps already learned such a lesson and not wanting to hinder their own development. Nor will disharmonic souls offer this gift—namely, your enemy will surely not do this for you. Only those who really love the soul will consider giving such a gift. As these souls enter into a new victim-victimizer scenario it is with full realization that the karmic bonds between them will become tighter. Such mutual karmic interaction leads to an almost inseparable relationship lasting many lifetimes. Only beings really loving each other would agree to such bonding.

Present-life victims, realizing the power and control they wielded and the suffering they caused in earlier lives, may see their situation from a new and higher standpoint. They become transformed once they realize the deep love on the spiritual level existing between those beings accompanying them in this karmic dance. This insight frees them, moving aside the barriers in the heart, allowing love to flow out, releasing them to the insight that they had chosen these experiences for growth— not punishment. They are not damaged after all, and now they are free. Once they see the hologram and enhance the responsibility for the choice, these magnetic messages inviting victimization simply dissolve. Renewed energy, flowing unhampered, quickly allows creative, positive responses to life and all its adventures.

It is possible to cast off the victim role in this life and become a creator, instead of originating a new victim role. When one realizes that he has already exercised power, demonstrated special talents or capabilities, an enduring self-trust is acquired. This new self-trust releases the old impressions of guilt, prejudice and fear, feelings that only corroborated our incompetencies, even bringing them into action. We may have habitually thought that we wouldn't make it or weren't worthy of being loved. Our new recognitions will release such thoughts. We begin to orient ourselves toward the finer, energy-rich frequency of real creative power.

Once we resonate harmoniously with the new reality that we are creators, we'll be transformed forever! We don't have to force our thinking into a new pattern or tell ourselves that since we now know the truth, we will do such and such with our lives. This transformation occurs naturally on its own.

In the sessions, one layer of consciousness after another peels away. We realize that we have possibly been healers, philanthropists, creators or artists. Without striving for it or thinking twice about it, that positive energy we were filled with in sessions activates us in new ways.

As we release ourselves from the residue of such experiences, we return to that essence we were before: energy. This pure energy is then again at our disposal—all is energy, the entire cosmos is energy, as well as the "emotional residue," which hinders us with its rigid energy formations. With this transformation the body becomes more open, more receptive, lays aside limitations, self-denial and deprivation, instead absorbing creative strength. We can then radiate this pure creative energy in many forms. That is a synopsis of the process of transformation from being a pitiful, embittered and helpless victim to a radiant, blossoming, creative being. The soul manifests itself, choosing to be creator and creative. Without another thought, we begin to radiate life energy and love, feel other people attracted to us, and become loved.

The cosmic pulsation of energy exchange, which flows out of our body and floods back in again, represents basic universal energy laws. It is a form of cosmic clockwork. Once this process is set in motion, wonders can occur. For example, we may only need to think of someone we'd enjoy being with and he or she calls. We can concentrate on a question and intuitively know the answer immediately, or it comes to us in wonderful "coincidental" ways through a newspaper article, a book or a comment made to us. We begin to understand that we're a part of the cosmic dance. Balance becomes an expression of energy, which we have the tools to comprehend and orchestrate ourselves: we want to for we are no longer on "automatic pilot."

Thus we begin to lay aside subconscious influences. We begin to break through the isolation of our persona—our mask. We release ourselves from what has been our unrelenting and apparently inescapable karma. We transcend the barrier of time as we step outside of the cosmic treadmill to become conscious participators in the divine plan as co-creators.

People coming to the Light Institute have one thing in common: they want to know their inner selves and commune with their Higher Selves. Many are very successful, famous, influential, and so many are astoundingly young. They all search for Enlightenment. They are longing to find a place in the world on a global level.

When Shirley MacLaine came to the Light Institute she wanted to learn more about the meaning of her relationship with her parents. This was very interesting because she presented a wonderful opportunity to demonstrate that we seek out our parents. It is one of the most profound insights we can gain in our spiritual development, as well as in the fulfillment of our worldly goals with purpose and love. In her *Dancing in the Light,* Shirley very vividly describes her urgent efforts to understand this subject. If everyone could learn to understand, as Shirley did, that parents are the first prototypes we experience in this three-dimensional earth-level reality, and that we chose our parents not for self-punishment or self-restriction but to hone certain capabilities, it would totally revolutionize relationships in this world!

I was, of course, electrified when Shirley came to tackle this delicate subject, afire with my vision of further research in this direction. This was the chance to free more people from false projections due to unclarified parent-child relationships. Whether we rebelled against these first archetypes or tried to imitate them in order to grow (whether we rejected or adapted to the relationship), we have the opportunity to really know ourselves when we view parent-child relationships in this new light.

C. G. Jung said, "Ask a man how he feels about his mother and he'll tell you how he treats his wife." My experience confirms this. Our initial experiences in this world concern who we are, what we can give, and how much strength we have in relationships. The answers within the archetypal relationships between parents and children determine the framework in which some of us are entrapped throughout our entire lives.

Nevertheless, if we can understand that we knowingly and willingly chose these parents from a higher level of insight and strength, we can recognize and work on the life themes that our soul prearranged for our highest possible evolvement. We can approach self-realization much more easily when we don't fight our chosen lessons, but accept and recognize that we are in the middle of a meaningful learning phase. The people with whom we struggle—the alleged opponent, the villain or the powerful authority—are recognized as great "teachers" whom we embrace. Wellsprings of energy become activated in us, new energy and levels of consciousness are opened up to us, and our paths of development unfold on their own. We really experience and understand this, not just intellectually: we become creative—creators. We realize that we can freely decide to do what is best for us.

More important, the emotional body becomes recharged. From the spiritual level we can relay to the emotional body that an experience, a cycle, a lesson, is finished! At the emotional body level we can completely understand that we were both victimizer and victim, that all roles have been dramatized, that this cycle is complete. Thereby we can step off that treadmill, that spiral. Actually it is not a single closed circle, not a dead end, but more like an ever higher and stronger upward winding spiral.

Because of new spiritual experiences and expanded consciousness at the spiritual level, the emotional body can also realize that it need no longer follow the reflection mirrored by parents and established roles, but can develop itself further. Through this new spiritual energy the emotional body is released of old emotional residues and is simultaneously capable of opening itself to a new energy, with which, with great fascination and joy, it will pull itself continually upward. The emotional body discontinues an old, boring and tiresome dance, thrusting itself forward into a new dance, a dance in the spiral of creativity.

MANIFESTATION
IN AFRICA

7

Realities at the level of "Creator" are adventures on the highest octave of human potential—to sit in the center of manifestation is to participate in miracles. They are there to remind us who we are. They force the consciousness out of time-space patterns and into the energetic cosmic laws of synchronicity. Time becomes the eternal pulse of energy as it ebbs and flows, out and in. It swells and gathers into itself all the creative components that then suddenly rush together in synergistic union—the birth of an event, a thought, a being. Manifestation is simply the visible display of this event, mirrored across the veil by its naked energy body.

We are here to learn, to remember the mechanics of miracles, so that we can perform them. It is the true purpose of our life—to create miracles just as did our forefathers, Jesus the Christ, Buddha and all the great creators of the generation of man.

Manifestations go on all the time in our world. We don't notice them because of our singularity of mind. If we begin to probe synchronicity, we realize that it is not time that is the great lord of reality, but it is the synergistic focusing of divine will. We can learn this art of focus; indeed, it is the only true education of our being.

Let me tell you one of my favorite stories of manifestation, as it exemplifies so clearly the unfathomable mystery of multidimensionality and at the same time its everyday appropriateness in our world. One day, as I held the head of a client— as I do when beginning a session—I began sensing the flowing of currents in my body that accompanies the expansion of consciousness. I suddenly received a telepathic message, more like an order, saying, "You must bring the Hopi to Africa to make rain." This command came from my Higher Self and pressed itself upon me in an urgent manner.

What an incredible idea! I thought. The Hopi know the art of rainmaking. For over four thousand years they have grown corn in the barren Arizona desert. As do other Indian tribes in North America, they have rainmaking rituals specifically for creating climatic and weather conditions harmoniously with Mother Nature.

I had had some experiences with rain myself. In the summer school we occasionally went outdoors on sunny, cloudless days to call rain with the children. We'd make noise with pots, pans and spoons, singing for clouds and rain to come and soothe us. During those six years it never once failed to rain whenever the children called for it. There is a Hopi saying, "If the heart is pure, it will rain." The hearts of children are pure and know no limits. If one tells them, "Now we will all go out, make noise and call for some rain," it will rain.

Therefore the telepathic message didn't appear totally impossible. However, I knew no Hopi people personally, since their reservations were rather distant from Galisteo. Not to mention that the sheer cost of flying to Africa on such a mission was itself an insurmountable block.

However, I accepted the message with agreement and joy, since it offered a wonderful opportunity to be present at such an event. It seemed a simple and natural way to help these people, instead of taking part in the continuous bickering between governments and bureaucratic organizations. People starve because their fields lack rain. To call rain seemed to be the best and simplest solution: bring them what they need the most. The magnificence of this thought overcame me in a flash.

Three days later I received a phone call from a client in Texas. She had to cancel her sessions but wanted her cousin Larry to take her place. He was an artist working with Indian motifs, and he arrived presenting me with a very lovely feather sculpture. He explained that many years ago he had been adopted into a Hopi Indian tribe by a now eighty-two-year-old Indian woman who was the wife of the last sun chief of the Hopis, and a daughter of a chief herself. She and her brother live in a very remote Hopi settlement high up on a mesa above the stretch of their land. There is no electricity there, no modern conveniences. All governmental offers of modernization were rejected, as they had witnessed the effects of change of lifestyle on their culture. They wanted to maintain the traditional Hopi way of living. Ever since they were children, they had participated in the cyclical ceremonies devoted to the blue Hopi corn. Blue corn is a staple in the Hopi diet, and considered a sacred food. For over four thousand years this blue corn has been cultivated without adulterations and without crossbreeding with any other types of corn. For generations the seeds have been collected, preserved, and planted. The rain would then be called, the ripening of the corn attended to, and this cycle of fruitation, seeding and planting repeated again and again.

When Larry told me his story, I related my Africa message to him. It so happened that he was on his way to visit his adoptive grandmother on the Hopi reservation. He was willing to relay my vision to her. A synchronicity was unfolding. The Hopis would soon enter their kivas, underground religious ceremonial chambers. Each year at a designated time they entered the kivas to consecrate prayer feathers that would be used in rainmaking. His arrival at the Hopi mesa would coincide with the prayers for rain.

He told the Hopis about the Africa idea, about how enduring drought caused such hunger there. The Hopis compassionately prepared prayer feathers for Africa as well as for their own land. The brother of Caroline, Larry's Hopi grandmother, was a medicine man. He said he could not leave Hopi soil, the land under his protection and guidance. He decided that Caroline would go to Africa to conduct the rain ceremonies. He instructed Caroline and Larry in the very specific tasks that needed to be carried out perfectly. This ceremony was to be a dramatic first. Never before had a woman conducted the rain rituals. He appointed Caroline to bring rain to the Africans in the name of the Hopi people. We all began a time of spiritual focus and fasting to purify our hearts for the rain rituals.

There were still a multitude of preparations for the journey. Our trip would cost between $12,000 and $15,000! Caroline, Larry, I, and my youngest child, Bapu, were to travel. I had borne Bapu in the ocean and the Hopi had meanwhile given him another name, bringing with it special gifts: Paloloca, meaning "water snake." The Hopis said his presence at the rain ceremony would have a special meaning. But we still had the problem of how to "manifest" the money, since neither Larry nor I had any.

I contacted an organization that had helped Africa by collecting millions of dollars from rock concerts. They said of Caroline, "We wouldn't know what to do with her." I tried to explain that they only needed to get her to Africa. Once there she would independently carry out her manner of bringing the rain by herself. This was inconceivable to them, since it didn't fit into their help scheme via food distribution. They elected to send latrines. Our conversation made the different conceptions of solutions for problems interestingly clear. Real solutions are like blades of grass: they look very simple and unpretentious, and are in certain ways the most wonderful and all-encompassing creations. I sadly reflected on my many years in the Peace Corps, when I had witnessed so many ill-fated attempts to help people because there was too little cross-cultural understanding.

I meditated daily and visualized us all going to Africa, although I had no idea how it would happen, and regardless of the fact that I knew no person or group that would participate monetarily, although I asked several.

Larry's cousin, who had given her session to him, heard about our project and decided to give the money for it. She gave us the $15,000 without tax benefit or any other advantages from it. She simply felt a part of the energy vortex that had developed around this project and wanted to participate.

In early May 1985, we finally departed for Africa. We traveled to Somalia because it was politically impossible to fly into Ethiopia, where the need for rain was greatest. We chose Somalia, lying on the East African coast, in order to bring the rain westward from there through Ethiopia, the Sudan and farther west.

In Somalia many people helped us, appropriating the vehicles, making the contacts needed, and finding housing for us. We drove into an area occupied by agricultural experimental stations and were greeted by the town fathers in a small village surrounded by corn fields. There we were in the searing African heat in a far-removed settlement with just a few straw-and-mud huts. Caroline was fully outfitted in the traditional Hopi medicine costume: black, red and green patterned wool blanket; a white skirt; and turquoise bracelets that radiated a special energy as "sky stones," along with all the other symbols of power belonging to the Hopi people.

We continued our fast after our arrival in Africa and kept to our special fasting diet of small amounts of certain vegetables and the blue Hopi corn. I was still breast-feeding Bapu, and happily his first food in New Mexico had been atole, a puree of this corn. We only needed to stir in water and drink the atole. We fasted for almost three weeks in this way before the rain ceremony began.

The night before the ceremony, Caroline laid out her prayer feathers and the ceremonial objects imbued with power. The Hopis call these feathers "Pahos." Caroline always wore high boots because of leg injuries encountered as a girl when a car ran over her. Despite her painful arthritis, she managed to keep her burdens at bay during the entire trip. On this pre-ceremonial night, she sat upright the whole night, without slumping, in complete quiet and stillness, without even once moving or stretching out her legs. I knew through my body work how much pain she must have had in her legs. Yet she remained sitting, still and silent as a stone, sunk into her prayer for rain, aligned with the purity of her heart and the gift, which was only possible with a flawless inner self-control. As a delegate of her people, she was here to bring their gift of rain to this drought-stricken land.

The next morning we went to the uncultivated field with the town fathers. We had brought about forty pounds of blue-corn seeds. Caroline herself had extracted every one of the kernels from the cobs that had been set aside for Africa. Caroline and I, the only females among a circle of men, sat on the ground, and she began to address the Africans. In the beginning it was apparent that the Africans found all this a bit strange—an Indian woman decorated with mysterious objects and accompanied by prayer feathers and odd implements. She said, "I bring this sacred corn from my people to your people so that you will know that you need never hunger." She lifted up a special fan of eagle feathers. Eagles are considered carriers of unusual powers by most North American Indians. She explained that she wanted to call rain to demonstrate that such a thing was possible. She told how this beautiful deep-violet-colored corn has been the unadulterated staple for the Hopis for over four thousand years. Astonishment replaced the grins on the Africans' faces as they expressed their admiration for a food passed down through the generations for four thousand years. Although they were little prepared or informed as to what Caroline intended, they were fascinated by the power of her words.

She spoke about the cycle of life, birth and death. She advised the Africans to sing to their future corn plants when the seeds were sown. She told them that they must express love to their plants and in their fields. The smiles in their eyes and shining faces showed that they understood what Caroline meant. Perhaps they knew from experience how important love, tenderness and singing are for every new growing life. This mutual understanding created a bond between Caroline and the Africans.

This beautiful, magical woman, standing barely five feet high, arose after her speech. Raising her feathers she voiced her prayers to the east, south, west and north, to father sky and mother earth, to all powers to bring the energy here. Then she poked the eagle feathers in the ground where she had created her shrine. She strewed yellow cornmeal around, as a carrier of the prayer powers. At the last she held her eagle-feather fan high up to the heavens.

Since our arrival the sky had been blue, but now it darkened unbelievably fast, clouds gathered and the first raindrops began to fall. At that point she began to plant corn seeds in the earth with a stick. The seed holes were a foot deep, much deeper than normal. With seeds sown this deeply one can't normally expect any germination within the first two weeks. In an amusingly authoritative style she instructed the town fathers exactly how they should sow the corn. She planted an area around the shrine and the men marked it off with string and bits of cloth. It began to rain so hard that we couldn't navigate our Jeep to our quarters on the coast. We remained overnight in a neighboring village, in which there were two round huts with cement floors.

According to the Hopi, one must return every day for three days to the prayer site to bless the area. On each of the first two days we were unable to return to the field because of the heavy rainfall. On the second day the Jeep stuck in the mud and we returned part of the way on foot. Caroline stood on the edge of the road and directed her prayers toward the shrine. On the third day after an arduous drive, we arrived near enough to the field to manage the rest by foot. To my surprise I saw the first green corn sprouts around the shrine. In three days they had germinated and were already about an inch high! It was a miracle. These tender green shoots stood everywhere, each with two leaves on either side of the stem. Inexplicable by any other means than divine grace!

There was no logical explanation for corn seeds planted so deeply to germinate and shoot up within three days. This miracle demonstrated that we have the choice to participate with Mother Nature if we focus our energy to align with her. It was a humbling experience.

At our departure, Caroline stood for the last time at the prayer site to send her blessings along the westward path in which she had directed the rain. She threw her cornmeal in this direction, as did we all, even Bapu, who had just turned nine months old. Bapu instinctively knew what it was all about and reached for the cornmeal that Larry was holding in his hand. Then he too threw the cornmeal in a westward direction, adding his love and blessings. We left Africa.

Later we were able to gather meteorological data from various international control stations, corroborating the event in a wonderful way. What they reported was this: As the rain began over the field, a mysterious storm brewed out of a direction from a coastal section of East Somalia. Never before had the rains come from that direction. The local people also acknowledged this as the gift Caroline had brought them. For unexplainable reasons this storm appeared and moved across Somalia and Ethiopia toward Sudan and farther westward. It brought rain amid the longest drought of that season. This rain returned throughout the next three months. According to the foreign minister, this section of Africa had its best rain in seven years. It wasn't a catastrophic rain, washing away topsoil, but a productive one, falling exactly everywhere Caroline had directed.

Unfortunately we have no further information on how the corn developed in the meantime, although we did give huge sackfuls to be distributed by an agricultural organization active in that area. The blue Hopi corn is especially suited for such barren regions, since it needs little water, grows in sterile soil, and produces such a nourishing grain. In the meantime a similar project is in progress. "Historical" drought-resistant seeds, cultivated for centuries by Indians, are to be brought to Africa with the motto, "The red man feeds the black world."

How can, how should, one understand the manifestations of prayer and power? Caroline was one with the elements of Nature, merging with the creative impulse moving Nature. She opened up conscious energy in other dimensions. None of us is actually separated from the world or the powers moving it. We can all participate in the manifestation and intentional releasing of these powers. We are composed of the same elements as everything else in Nature. Our bodies are approximately 80 percent water and our blood approximates sea water. As do insects, birds and other animals, we too can activate that sphere in our brains capable of understanding weather influences.

If we want to manifest something ourselves, we have to tune in to higher octaves. The Hopi accomplish this with their rituals, which focus their attention and intention. Other indigenous cultures in the world also do it. Only we have lost contact with these levels. If we regain this contact and awaken those slumbering spheres of our brain, we could achieve the natural resonance necessary to participate with Nature on the manifesting level.

Working with weather patterns is a wonderful way to start exercising our capacity to attune to the multidimensional world. The results are immediate as Mother Nature is very willing with her energy. Our imbalances, caused mainly by linear and therefore inaccurate or incomplete information, could also, of course, create disaster. Nevertheless, I feel certain that this is the answer to pollution, radiation and drought.

The attunement aspect is so critical here and yet, as we ask for and receive holographic answers, we could learn to speak the cosmic language of energy. Every action or event imprints itself in the latticework of reality—not on a time continuum, but on an interacting point of synergy, which shapes matter in all directions, past, present and future, as well as echoing through the energy layers of dimensions. Thus, any thought, such as a question, undulates out across the waves until it connects to its counterpart—the answer—which undulates or images back via manifestation. Once we become aware of this "circular arcing" relationship of cause and effect, we will become very sacred with our questions, our projections, our desires.

It is crucial to recognize the difference between manifestation on the level of cosmic synergy and the octaves of manifestation by the force of personal will. Beams of willpower can be focused for an impregnation, i.e., that a certain situation comes into being. This is done by the projection of a feeling, or a sensory experience, which should lead to a strengthening of intention. We can all do this form of alchemy, but if it is done through personal will and desire to override something, it opens up avenues of karma.

There is a difference between our earlier applications of alchemy, in which we established personal power to force our will by magic upon nature or other people, and what I described about Caroline. On one side is the striving for control, the exertion of powers over natural laws, for example, converting iron into gold. Alchemy uses personal power to force our will by magic. On the other side, the consciousness opens and expands, nothing is separated from us. This opening allows us to receive knowledge and to pull in the natural energy flow of creative synchronicity in which the "will" belongs to the universal flow, not to personal power!

Dowsing is another wonderful exercise of attunement to the natural world. It is a very tangible, physical aspect of our inner consciousness power, which has potential access to all levels of awareness.

I'll relate a small example from recent times. The ability to find underground waterstreams or lost objects requires no special talent and is not something limited to a chosen few. Everyone can develop such capabilities, at least to a certain degree. The children had learned how to use a divining rod— all except the very small children, who were unable to hold a rod or pendulum properly.

They were always delighted when an invisible energy suddenly pulled their divining rod downward. This is a form of dowsing, in which we use the electromagnetic fluids of the body to attune to and access all levels of awareness. It's as if the fluids themselves are the receptors that translate that which sits within the void of the formless into the manifested reality. In other words, the "sloshing" of the fluids resonates with the knowing and selects the "yes" or "no." This is a very tangible, physical aspect of our inner conscious power, which has potential access to all levels of awareness.

One of the integral parts of the Light Institute process is cranial work, which focuses on balancing energies in the head. The master glands, the cerebral spinal fluid, as well as the intricate relationships of the bones and sutures, are all realigned. This is dowsing on its highest octave. Holding a head and perceiving the various different fluids moving through it demands acute sensitivity. I had been working with David and Richard, who do the cranial work at the Institute, to fine-tune their dowsing sensibilities. We decided to practice these abilities by participating in an unsolved mystery.

In the mountains near Santa Fe a young man had recently died, although no one had yet found his body. We obtained topographical maps and went to work with our pendulums to search for the body. David and Richard were to ask inside where it was. They went into the mountains and were led by the pendulum marks on the maps directly to a spot in which they found a skull. This turned out to be the skull of a murdered woman.

It is fascinating that since they were focused on heads as their dowsing frame of reference, they synergistically manifested a skull. Walking directly to that skull, lost in the Sangre de Cristo Mountains, was truly finding the proverbial needle in the haystack! We were all ecstatic! We forgot about looking for the young man who was later found at another spot on our map.

The wonderful thing about these kinds of "natural" manifestations is their practical application in our daily lives. For example, I once successfully maneuvered a water line from my well across a distance of about three blocks without hitting any other power or water lines. It would be wonderful if these skills were taught in schools. There is such a fulfilling joy in participating with nature that gives us a feeling that we truly belong here. As we begin to listen to the flow around us, we more easily listen to the flow within us. In such a state of heightened awareness, it is impossible for disease to sneak up on us. A short meditation each day, to clear pollution for example, could produce such resounding results as to entirely transform our perception of who we are—in our life and in the world.

Feelings and understanding are the boundaries of this world. The boundaries don't lie outside, in space, but rather within ourselves. If we begin consciously to participate in creative processes, our vision of developmental potential, free choice, responsible selection of future conditions and events, can be accessed through "circular arching" pathways. We will learn to realize that we actually can call rain, and much more.

Behind this conscious part of the creativity lies the "yin" energy, that magnificent "feminine" power, which is concealed in the vortex of the formless unmanifest.

This power is identical with the divine force. We are here on this earth experiencing a synchronistic activation and release of these creative powers very practically in our everyday lives. This feminine energy, this intuitive, formless power, this inner wisdom can now be experienced in fully new ways.

THE FEMALE
EXPLOSION

8

One of the most magnificent universal phenomena is that of synergy. Synergy is the perfect movement of energy, which seeks its own kind and gathers itself within a particular pulsation within a most divine timing, until that energy meets itself as critical mass. It builds up to a particular pitch and then explodes, creating a kind of fission from which something entirely new proceeds. There is a synergy operating now in which the flow of manifestation through the universe has been heavily weighted on the "yang," or masculine, side. And as that energy has rushed the universe, it has left a pulsating mass of its opposite, which has begun magnetically to gather itself and in its gathering has reached that pitch of critical mass that allows it to become manifest. That mass is the energy of the feminine. It is a *new* energy that is freshly palpable around the world. It is, in fact, an explosion of the female.

Perceiving synergistically is a powerful way to alter our perception of linear time and helps us deepen our capacity to experience multidimensionally. For example, the incubation of female energy has not been a matter of eons or centuries, or the passing of cultures within time and space, but an energetic expansion of a quality that enlivens itself and moves into its peak density and pitch so that all its energetic components are pressed together and translated into form and matter, into palpable presence. This is exactly what is taking place with the female energy that is in motion now, making itself felt because the more consciousness polarizes itself into one kind of energy, i.e. masculine, the more it creates the space for its opposite to come forth. Such are the laws of nature—to bring into balance all the extremes, to come back to homeostasis.

Chinese medicine offers us a wonderful example of the motion and the interplay between energies. It describes the masculine as yang energy, which is energy in motion, energy that thrusts or rushes out to build, to manifest form to visibility. The further that energy moves, the more profound the recognition of its connection back around the circle to the yin energy, which is the formless, female creative source energy that allows for the blueprint to be taken out and manifested by the yang energy. Thus we see these figure-eight patterns that move in and out of each other. On the one hand they seem so dualistic; the male, yang, so visible and forceful, and the yin, female, invisible and only subtly present. Their interaction delineates the balance of nature: one cannot exist without the other. The masculine energy has stretched itself out in such a pervading fashion throughout all our cultures that an activation, a synergistic awakening of the female energy is beginning which, through its presence, alters the very nature, intent and expression with the yang, masculine force. Within the human experience, masculine and feminine have been sharply delineated in terms of their respective manifestation or expression in virtually all cultures. Yet, the separation of these two energies from each other has created the very impulse for nature to shake itself free of the stagnation of limited expression, to create a new order, a new balance in the world.

The structure of our religions is a classic example of the yang order in the universe. Because churches have been presided over predominantly by the male form, they have perpetrated separation by not allowing the female active participation as vehicles of divine knowing. As long as the mental body is imbalanced in this way, the teachings of any church will tend toward dogma, separation and ultimately war because it is missing the loving, compassionate energy of the female. It is interesting to contemplate that the spiritual source of the world's great religions—such as Christ, Buddha, and so many of the Indian saints themselves—were in complete harmony with the male and the female. They magnificently demonstrated the yin energy within themselves and used it as a tool to heal, to nurture, to inspire all those around them. Yet, when their knowledge was delineated, crystallized, categorized by the yang energies, the dogmatic perception actually created the antithesis of those profound messages. Though most organized religions are more than reluctant to allow women to participate in a place of leadership, female spiritual leaders are beginning to surface all over the world. It is breathtaking to sit in the presence of such an illuminated being as Gurumayi, who is the head of the ancient Siddha lineage (those who have the power to awaken the kundalini energy in the body, to give "shaktipat"), and who so perfectly exemplifies to the world a model of balance and love and radiance. She inspires us to seek out our female energy and trust it to bring us into heartfelt harmony. The explosion of the female belongs to us all. We can, male and female alike, embrace this evolutionary energy, which allows our consciousness to soar and our human experience to be transformed.

We carry within us the seed of experience wherein the male and female merge. Even on the earth plane, until the experimentation in Atlantis separated the male and female energies, we experienced those energies as one. In many other dimensions of reality, life expression is androgynous, male and female merged, and we carry those multidimensional memories within us as well. The Atlantians, in manipulating the genetic code, i.e., separating the male and female, began an experimentation in which these distinct qualities separate from each other could be more fully understood. Since the male, yang energy is dominant in a structural modality, the experience of that energy has very clearly left is mark on the world. In all aspects of our physical and subtle bodies, the yang energy has been at the forefront. Emotionally, we learn to use anger, which is yang energy, to suppress or control fear, which is an underlying yin energy. The anger inherent in separation created the illusion of the struggle to survive. Emotionally, mentally and physically, we became obsessed with elements of survival, manipulating matter, forcing personal will upon the natural world, and focusing our attention on personal power. This further separates us from any hope of merging. It forces us to lean upon the fight-or-flight reality wherein only one of us— you or I—is the victor, the king of the mountain. Although we think of time as being an endless continuum, if we look at life as energy instead of through the linear measurement of time, we see that the energy of this yang reality is finite, limited. It comes to an end and dries up because it is cut off from its arc, its bridge to the feminine, which is the source of its creation.

Unless the motion is indeed circular, the spiraling element of evolution does not occur synergistically and we go into a dissolving, destructive, deleting phase. Yet, because life is the plaything of the god-force, nature shifts itself and rebalances. The female, yin energy begins to make itself felt, begins to whisper within the memory, within the infrastructure. The female, though it carries the blueprint of possibility, is within the realm of the formless; of spirit, of knowing, intuition— the designer of the possibilities, which leans upon the active energy of the yang to bring those possibilities into form.

This is the mechanism of conception—the threshold through which life expresses itself in our world. If we begin to study the energetic quality of conception as being a synergetic point of reality, we can access miraculous universal power to create in ways unimagined. The laws and principles of conception, wherein the unmanifest becomes manifest, can be utilized when the male and the female are in sync not only to reproduce in kind our own energy but also to produce, with the abundance of creative genius, any form of life, whether it's a flower or a feeling. At the same time we could dissolve back through the threshold into unmanifest any material, crystallized form that is not harmonious, such as pollution or hate.

The explosion that is taking place because the female energy has concentrated itself and reached critical mass simply means that the formless, concealed from our view, is suddenly surfacing. It takes on a living form, emptying from a whirlpool of darkness and unconsciousness. It manifests in this world, urging us to take notice of its inherent energy, prompting us, regardless of our gender, to contact it, accept it, and creatively use it. One of the things that will most heal the separation between the male and the female is an inherent recognition that these energies are contingent upon each other and that all form contains them both. Women have masculine energy within them, as men have feminine energy. Our consciousness can guide the balancing of those two energies to produce the highest octave of awareness that allows us to utilize them in sync with each other and change the way we live in this world.

The great knowing, these great intuitive gifts inherent in the female can very practically and pragmatically enhance the synchronistic success of the yang energy in its pursuit of manifestation. For example: the most successful executives and manifestors always utilize the intuitive, "female" capacities to attune themselves to an expanded level of consciousness that guides them to clear decisions. By the same token, the woman who frees herself from the unexpressed knowing, the invisibility and darkness of the holder of the form, can give great gifts to herself, her children, the world around her. Sometimes, because the emotional body has a repertoire of such great pain and caution, it develops its defense armour to such a degree that it actually forgets it can access both of these energies freely. Thus it experiences profound emotional separation from its opposite and an underlying longing that cannot be quenched. Men and women create so much pain between them because of this exhausting search for the missing energy—we seldom experience it directly, we create profound illusions as to the qualities of a "real" male or a "real" female.

Perhaps the most positive angle of divorce as a growth tool is its power to disengage us from the lethargy of projection. When we live together, all the cultural as well as unconscious imprints of male/female roles come into play. Our entire communicative interaction is based on the assumption of those opposite positions, and we therefore invest heavily in the idea of our mate representing to us what is all too often a caricature of what is a male or female. When divorce ends the illusion that "male" or "female" is outside ourselves, we begin to discover it within.

The alteration of roles is especially poignant for divorced parents; the mother suddenly must learn to play the father, and in the case where the man has custody of his children, learning to nurture and surrender to the ever-present needs of the child becomes a necessity. Most commonly, the children stay with the mother and so they begin to search for the father, or yang energy, within them. In terms of the male/female evolution, it is fascinating that the yin must endeavor to absorb the yang as we move from a predominantly male-oriented world to one in which inner wisdom will guide the path. We must understand that the dissolving of a male role model outside us indicates that we are ready and able to bring forth the one hidden deep within. Our Higher Self designs the experience like a graduation—it does not *ever* happen until we are ready! This understanding is very healing to children who have lost their father either through divorce or death. It brightens their hearts and brings back a sense of power—they feel they are strong enough, or masterful enough, not to need the external model. It relieves them from much guilt and fear to experience it—not as separation or abandonment but as an initiation of strength, which they have already passed.

Through our sessions at the Light Institute, we bring them into conscious experience and contact with the energies they must now find within themselves. They become ecstatic, bringing to this new expression great compassion and blending of the male/female energies.

Delightful rewards come to us when we explore our multi-incarnational realities, i.e., past lives. Contacting those opposites within our repertoire of experience really enriches us. In other words, we have all been both male and female. Sometimes we identify ourselves strongly with one polarity or the other and it is a great experience of growth to reaccess within ourselves our opposite. When a woman reexperiences herself as a male, with all the urges and compulsions to manifest that are inherent in the male, she more easily activates her dreams. When a man discovers that he has been a female, the tender, nurturing, creative energy begins to flow without resistance, because he can then recognize it as an energy that belongs to him!

One of the greatest mysteries in life is the mystery of procreation, of giving birth, an innate power given only to the female (in human realities). Because men are cut off from experiencing birthing directly, they perpetrate their own fear of the female, the remembering of being separated by birth, rather than the sacredness of *giving* birth. Thus they replay the polarity created by that separation.

Birthing is the greatest of all initiations in the art of surrender. Once the birthing process begins, there is absolutely no way out. This teaches us to surrender ourselves to something greater. The subtle remembrance that we are in relationship with a universal force helps the female to attune to divine energy. As she surrenders and the birth sequence plays itself out, another profound truth emerges. Surrender brings not death or inability to survive but, always, the birth of something new. It is in the movement, the "moving on," that we survive. Our bodies understand this—it is what makes the desire to conceive, to reproduce. Only our egos, lost in separation, have forgotten. Men who can experience birth through their cellular memories come back into touch with the closure of the surrender circle—i.e., birth brings a recognition of great strength and power; we surrender and everything changes. We are part of that great flow.

By becoming conscious of the necessity to relate intimately with both aspects of our being, we can create new realities that optimize experiences of both sexes heretofore in the domain of the opposite. For example, when there is a birthing that takes place, a man can bond with the creation of a new being. As men experience themselves giving birth in sessions, it literally breaks open the deep recesses in their hearts for loving and nurturing. They come away from the experience with a new look in their eye—that look truly transforms how they see themselves in their lives! This kind of personal illumination can bring forth entirely new societies of evolutionary consciousness.

Birth is a very powerful initiation, and initiations have always been designed to allow us to recognize our potentiality, our mastership of any octave. I myself have passed through the funnel of this birthing initiation six times. They were all experiences of natural childbirth. The fifth birth, of Teo, I orchestrated myself, with my children and several friends around me. I will never forget the sensation of cutting the cord myself. It was a powerful act, bringing my beloved Teo into the world and receiving her with my own hands and then cutting her free again!

The circumstances of my sixth pregnancy were so mystical that I realized from the beginning it would need to be a totally new kind of birthing experience. I was four months pregnant before I realized what had happened. My first three periods came normally and on time. There were no symptoms of conception at all.

Two months prior to the conception, I began having light-body experiences in which my consciousness became aware of a light body that would lift up and move independent of my physical body. I am not speaking of an astral body, but a very light frequency that created very strange sensations in my head and gave me a new octave of information. These experiences often left me feeling lifeless and in a heap on the floor. So later, when there was the customary tiredness of pregnancy, I didn't recognize it because I was still engaged in the light-body experience. Everything about this final birthing initiation had the mark of something extraordinary. I wanted it to be the most beautiful and perfect birth—as do surely all expectant mothers. I wanted to return to the ecstasy of my first delivery, with the firm intention to use this birth as an instrument of transformation for my own life.

In Chinese medicine it is said that the female has the chance, through pregnancy and birth, to purify herself of all old deposits, impediments and cellular wastes. With the elimination of old energies she can be filled with new energy. (Perhaps this is one reason why women live longer than men.) By this process women not only give birth to another but also become "reborn" themselves. At forty-two, I was ready to be reborn.

I didn't just intend to cleanse myself of old energies, as the Chinese described, but to completely change myself. I wanted to consciously experience the symbolic meaning of birth on all levels.

I discovered that Russia had developed a technique for birthing underwater. The information I collected on this was encouraging. Research there had shown that the fine brain cells, which develop during the last six weeks of pregnancy, were not disturbed by underwater birth, in contrast to "normal" deliveries in which the oxygen crisis tends to destroy them. These are the delicate brain cells that involve telepathy and higher brain functions. Secure in this knowledge, I decided that my sixth child would be brought into the world underwater.

The sea symbolizes cosmic consciousness and is the mother of all creativity; it appeared so appropriate to bring a new being into the world there. The amniotic fluid surrounding the fetus is like sea water. By being born in the sea, the baby is not immediately thrust into a cold, hostile environment of air and atmospheric pressure, disturbing the delicate brain cells and causing birth trauma. Instead the infant exits into a large waterbath resembling the environment that nurtured it during the first nine months of its life.

The being coming through me must have sensed the need to become a part of the ocean environment. He allowed us a whole month to acclimate ourselves to our tiny Bahamian island, during which time we all meshed our consciousness with the life of the sea. Every day I swam about a mile around the island, becoming familiar with the individual fish that inhabited certain reefs. I grew to trust the water and myself as a part of it. Everyone in the family had a specific task to perform at the birth and was well rehearsed as to what they could expect to witness. I made it very clear that I was taking responsibility for all my choices and that any emergency situations were mine to handle. I was forty-two, and because this was my sixth birth, I was considered a high risk for home delivery professionals, and knowing that I must be completely responsible for myself and this new life actually gave me a very exalted feeling. My body would perform this task perfectly because it had to, and it knew how. The depth of my communication with my body was astounding: I experienced intimate dialogues on subcellular levels, clear commands and feedback! I actually knew exactly how the birth would go— I saw it, and it did. I was inwardly and outwardly prepared by the experience of my previous deliveries as well as the many births I had assisted in the Peace Corps.

My eldest daughter, Karin, then nineteen, was my labor coach, breathing with me in the last rush of intensity. The baby's father, Tuss, applied pressure during the contractions and provided a human chair. My friend Richard brought his peaceful loving energy to the event by playing the flute and taking pictures to be forever cherished. My son Britt, fifteen, watched intently throughout the process and circled around on the side of the sea to protect me from any intruders. Megan, twelve, cut the cord with professional precision, as she had practiced, when the pulse had stopped. Teo, six, carried our new one onto land after our first blissful half hour in the sea. It has always been tradition in our family for the youngest child to be the first to receive and welcome the next new member, and Teo did this with a visible sense of honor and awe.

A great cosmic joke was played on me about this birth. In

the Peace Corps, I had seen native women giving birth in unusual ways. When I heard that the Maori Indians of New Zealand gave birth in the ocean, I totally accepted it as true and set about to follow in their path, feeling very comfortable in their wisdom. I would have gone to New Zealand but it was winter there, so I chose the Bahamas for its pure, warm waters. After the birth I discovered that it was a Maori myth that the gods came down and gave birth in the ocean. I laughed with the tears rolling down my cheeks until I could no longer breathe. I had inadvertently become the first woman to give birth in the ocean—while I thought I was following someone else! I profoundly thanked my Higher Self for the trick, as it allowed me freedom from the slightest doubt while I surrendered myself totally to this breathtaking cosmic experience, completely devoid of fear.

I purposefully chose the powerful, symbolic environment of the sea to heighten my experiential reference point to the cosmic level, rather than just the personal. I felt a profound longing to connect the innate energy of the female back to its ultimate source. My desire to embrace this physical experience within its own womb of the spiritual so quickened my perceptual capacity that it utterly changed my life. I had to use my yang energy to give birth in this way. I did not realize how, by allowing myself to manifest an impossible dream, I would become so fearless! I would become a part of an endless, timeless, universal flow. That experience caused me to release and expand my focus from self-perception as a passive being, unable to participate on creative levels, to an experience of myself as part of a cosmic energy so vast that survival was not even a question of concern. The reality was the ecstasy of merging, of becoming part of something so much greater, that

I actually had the experience of birthing my *own* self as a new kind of being!

In whatever way we explore ourselves, palpating, caressing, contacting our higher selves, our limitless, divine, universal spark, we can translate that energy by assimilating its experience. This completely alters our relationship to ourselves, and therefore, our relationship to the world. Before males and females can hope to create a new kind of relationship between them, a new way of being together, they must explore their own relationship to themselves. As we experience wholeness, we release the illusion of imperfection, of lack, and therefore, of longing and need. If the woman has found the man in herself energetically, she will open her capacity to manifest, to create form and be visible. When the man has discovered the woman in himself, he will open to the tenderness, the nurturing, the knowing. It will no longer be necessary to compete with one another, to limit or to outdo each other. If we no longer need each other, we will have no false expectations. We can cease projecting ourselves onto each other. We no longer need to oppose the "other." If there is no resistance, or illusion of separation, then we energetically, effortlessly begin to merge, and that merging releases tremendous amounts of energy that we can harness to create new ways of being in the world; new ways of being together. Our conscious pursuit of multidimensionality is a great tool to give us a frame of reference for these male and female energies within each and every one of us, so that we can perceive one another in an entirely different way.

It is an amazing experience to move through a past-life scenario in the opposite body than you have now. The sensations of being inside the body itself—posturing, making love—are truly illuminating. In many subtle ways this experience allows us to amplify our whole physical motor perspective. Once we release the body's memories, though we dislike the contents of a lifetime, the essence of all positive experience can be retained and utilized by the body. Many people report a major improvement in their sex life after our past-life sessions. The reason for this is that the body now has both the male and female point of view in terms of sensory facilities. This produces a fabulous blending: the self dissolves and we experience the aggressive, ecstatic frequencies that occur with merging.

We are at that point in life where the feminine manifests itself consciously and actively in the external world, where women also use yang power. Simultaneously, men discover something new in themselves underneath their strong shells. They discover the life force, sometimes called "shakti," which is the feminine creative kundalini power. Men discover something moving in themselves and begin to explore this. They search for tenderness and femininity. At the moment of conception the formlessness explodes into the form, and an animated but invisible power emerges in such a way that a totality is revealed in material form. This means that we consciously and willingly participate in the process of conception. We must rise to such levels of awareness and expand our consciousness so that we are able to utilize this divine energy of conception.

Feelings and understanding are the boundaries of this world. The boundaries don't lie outside, in space, but rather within ourselves. If we begin consciously to participate in creative processes, our vision of developmental potential, free choice, responsible selection of future conditions and events, can be accessed through these "circular arching" pathways. We will learn to realize that we actually can call rain, for instance, and much more.

Behind this conscious part of creativity lies the yin energy, that magnificent "feminine" power, which is concealed in the vortex of the formless unmanifest. It is waiting to be utilized by us all. We are here on this earth experiencing synergistic activation of divine force, which requires release of these creative powers very practically in our everyday lives. This feminine energy can now be experienced in new ways.

HONING
THE
ENERGY

9

We all can become manifestors in our lives using the intuitive guidance of the female within, whether we are men or women, to guide us into conscious synergy. We have only to embark upon the wonderful adventure of attuning our awareness to the different energies in play around us and consciously directing those energies for the purpose of harmonic manifestation. It's a skill we were born to learn.

Consciousness is not something that just magically happens to us. We must seek to become aware of things unseen, such as the whisperings of nature, which can be extremely important to us. Our bodies have unlimited perceptual potential that can be tapped and trained to carry us across the barriers, through the veils—into heightened awareness. The trick to wielding this intuitive knowing lies in honing the energies of all our multidimensional bodies to work in concert with each other. This creates an integration of the whole that allows us to be simple and clear even in the presence of vast informational energies.

If we are to perceive beyond our five senses, we must become aware of our own body's energy fields and how to heighten or quicken its energy. There is much to learn about how we interface with the world around us, how we respond biochemically and electromagnetically to the foods we eat, the drugs we take, and the emotional moods of the people around us that interact with our own.

Considering first our own energy field, we see that our body mirrors the same messages as our feelings and thoughts. In fact, if we do not express our feelings, the body will store them just as it stores biochemical substances such as glycogen. In other words, the body is a magnificent translator of matter to energy and energy to matter. It views emotions as energy and ultimately stores that energy as matter within its cells. If the emotional energy is not subsequently released, it will condense into matter and create blockages within the body itself—leading to disease.

The body is a fascinating road map of emotions. It literally deposits specific emotions in designated organs and areas that physically express and symbolize those emotions. For example, anger is stored in the liver and the TMJ joint of the jaw. The liver is considered by the Chinese as "the official in charge of the chi—or energy." Western medicine concurs with this in confirming that glycogen, which is a complex form of sugar that the body breaks down to produce quick energy, is stored in the liver. Anger triggers the sympathetic nervous system to fight or flight, requiring the liver to release its stores of energy to meet the immediate situation. Drinking alcohol also releases energy almost immediately into the bloodstream. We all know how easily someone who is drinking becomes angry. We can readily see anger expressed in a tightened jaw, which may hold the anger there, set against the world, awaiting the next explosive opportunity to express it, and then repeat it over and over again.

The left shoulder is where the body stores family conflict, those imbalances of familial responsibility that are karmic catalysts of growth. It is interesting that heart disorders commonly include pain in the left shoulder and arm. Clearly, disharmony between ourselves and our loved ones hurts our hearts! The body never lies. It holds the secrets of our innermost conflicts. We can learn to read it. Anytime the body calls our attention to some area, we can know that it is whispering to us about a congestion of energy. Even if we simply give it a color that directs our energy there, we will dissolve the imbalanced energy—whether emotional or physical in nature—before it congeals into matter and becomes disease.

By perceiving those radiating messages emanating from the body, we can consciously select the ones that truly enrich us the most and teach ourselves to focus only on those higher octaves that emit healthful, whole energy-wave messages to the world around us, so that that is what is mirrored back to us. If reactive feelings appear, we can go after their cause.

We can take a look at our physical energy system first— it is perhaps the most important to our consciousness because we experience it directly. How does this system react when the body is in motion or active in some sport or fitness training? To what degree do certain foods or drinks foster or hinder our performance? To what degree does clean, fresh air, or even our attitudes toward life, influence our well-being?

Our body is really the most wonderful energy hologram to help us understand how intrinsically interconnected we are with everything in the universe. It is a universal microcosm that truly interacts with the external macrocosm. Our entire nervous system, the brain, the cerebrospinal fluid, all interface with more subtle energy currents acting as conduits in and around the physical body. These energy currents flow through the meridians, the 72,000 nadis known to eastern masters, which are energy capsules like acupuncture points themselves, and the electromagnetic rays of the auric field revealing the innermost secrets of the physical, emotional and subtle bodies. We have only to become aware of them. People who are open and sensitive to the limitations of the auric field can, for example, pinpoint psychological and physiological problems from irregularities or colorations in the aura, as has been captured in Kirlian photographs and other electronic devices sensitive enough to record them.

How can we reach ecstasy? How can we be happy in daily life? Without a doubt a lot depends on the extent to which our bodies are in balance. If we let our body go, if we maltreat it or poison it, our emotional body won't react with joy to inner or outer impulses. Our emotional body won't be able to hold itself in balance because it is so biochemically inter-dependent with the physical body. If we as individuals don't live with joy and love, we will reflect chaos and spiritual confusion worldwide, because there is a profound energetic interdependency between the rhythms of individuals, societies, cultures and nations.

We must develop as much confidence in the energy system of our environment and planet, from which we are not separate, as in our own energy system and its rhythm. Sensitive children and adults very often perceive negative events happening around the world. Sometimes it is earthquakes or other natural dis-asters, and sometimes it is human disasters such as war and the emotional imprint of hate and revenge that ring out into the collective unconscious. These dense energies are usually experienced through the system, which registers queasiness and upset stomach, as well as undefined anxiety and depression. Children are famous for their frequent stomach troubles, which quite often are from sucking in, spongelike, the negative vi-brations around them.

Simply by becoming aware that our emotional antennae extend around the world, we can alleviate a lot of confusion within ourselves by checking to see if what our physical body is recording is about *us*. Take a few deep breaths and allow the mind to still itself. Then just ask, "Is this mine?" If you get the feeling or hear an answer that it is not, focus your attention on radiating light out the solar plexes. You can even ask "where" something is occurring, or what kind of thing, and receive messages and impressions that you will find to be quite accurate. Practicing this kind of global consciousness is very healing to our personal selves, because we begin to imprint that "we count!" If we want to positively develop the emotional body and experience its highest octaves—ecstasy, rapture, love—in our everyday life, in our family, in our work, within all that our life plan has created, then we must forge the energy of our body, so that it vibrates on the highest octave.

If we allow pollutants into our body, our energy system won't be able to vibrate at such high frequencies. The emotional body can operate in wonder and bliss only if the body functions at its best. Sugar imbalances create mood swings that are truly uncontrollable by our will to be "well behaved." We are using a physiological vehicle that reacts automatically to biochemical substance. If you insist on that alcoholic drink or that Coke, be prepared in two hours to experience a substantial change in energy as the body cries out for more quick sugar. If it doesn't get it, you will feel angry or depressed or exhausted—nobody is exempt! That emotional roller coaster puts great stress on the body, which is attempting to balance itself out. Our normal emotional states of fear, irritability, doubt and the like often manifest through purely bodily conditions. Physiological disharmony in the body due to ingestion of chemicals or damaging substances can lead to this. Certain foods, alcohol, drugs and air pollution overtax the liver, which is the organ of purification, and throw the entire body and nervous system into imbalance.

If we want to rise to higher octaves, we must purify ourselves! We begin by purifying the body. We must eliminate those poisons that cause us—not just physically but also spiritually—to remain, relatively speaking, on lower levels of consciousness. To pose a comparison, the faster a car is driven, the more critical it is that all parts function without friction. It is critical that even the smallest pressure variation on the tires be either avoided or balanced out.

Meditation is such a balancer for the body. If we meditate, we experience harmonious conditions of consciousness. Quiet rhythmical alpha waves neutralize disharmony in the body. (Alpha waves also occur during sleep.) These brain waves allow a regeneration and rejuvenation of the cells. We must begin to consciously ask ourselves: What does my body need now? Perhaps it requires a certain food or a change of diet to create a stable harmony that can increase in frequency without damage.

A small number of enlightened beings from all cultures have understood that by healing the body in certain ways, we can quicken its frequency to attain spiritual levels of full illumination. The practice of yoga establishes a heightened balance of the endocrine system that stabilizes the entire body. Breathing techniques have been used around the world to trigger the body into heightened states of awareness. Both yoga and breathing purify and rid the body of toxins automatically. Unencumbered by the need to pay attention on levels of pure survival, the body easily transcends all the lower octaves and moves into frequencies of light.

Normally we don't operate in a framework that allows us to move back and forth between these and other dimensions. Nonetheless, every time we experience a higher vibration, it creates a point of reference for our consciousness, making repetitions desirable and possible. Our body strives on its own to experience again and again these newly created realities.

We can assist our body—it is eventually entirely up to us. Very strict decisive rules and guidelines about what to do and what to avoid generally activate repercussions from our emotional body, which does not like control games of negation. Yet, if we let our body tell us, it will point out very specifically what helps it and what slows it down. It requires nothing more than a conscious attunement, a fine awareness of our energy system. Wine is, for example, very good for the digestion, especially a good old red wine that is unadulterated and unpolluted. But alcohol lowers the vibration! The body will demonstrate this very graphically to us if we tune in. If we wish to raise our sensitivity, to accelerate our inner vibrations, to gain access to higher dimensions, to grasp new realities and be able to activate them spontaneously, then we must stay away from alcohol.

We have observed that almost all people coming to the Light Institute change their habits of their own accord. They become more conscious of what they eat. Interestingly, no longer eating meat is one of the first changes the body often desires to make: people simply lose their taste for it without even knowing why. Consider for a moment: We no longer live in the good old times in which we, as the Indians did earlier, first established contact with the spirit in the deer, asked him to do us the honor and in specific spiritual attitudes let us take his body. Every part of this body was usefully employed, be it as food, leather skins, or bone instruments— always with the life-giving spirit in mind. Meat eaten in that way has truly a nourishing substance for the body. Where do we find something like this today? We keep cattle on pastures or even cooped up in stalls. We feed them growth hormones. We stuff them full of steroids and antibiotics. These chemicals are deposited in the animals' flesh, and then taken up in ours. Our system can discharge such chemicals only with great difficulty. After extensive research, it is undisputed that these substances have dangerous effects on the body. There is a biochemical side effect that is equally as treacherous to the emotional body.

Have you ever visited a slaughterhouse and seen how chickens or pigs or cows are commercially prepared and killed en masse? Chickens are slaughtered on conveyor belts, pigs electroshocked, cows are shot, each waiting his turn in line in knowing expectation. It has been scientifically substantiated that animals are just as capable of mutual experience and perception as are we. This means that every animal butchered for commercial consumption dies in fear. Bursts of adrenalin are released, the fight-or-flight reactions are activated. All this happens not only in the moment of death, but even before, while they share the horror of their fellow animals. The shock and fear work directly, physiologically on the body. The aforementioned substances are pumped throughout the body into the flesh. We consume this flesh, and thus we absorb their fear! We wonder why the world is full of fear and paranoia? We still wonder why fear principles dominate in the world? Fear is a compact, slow-moving vibration, a dull energy. Fear numbs. It spreads out like a membrane over our sensibilities until we no longer can perceive with any clarity at all. It becomes the all-encompassing angst that wears no face, has no name and yet whispers constantly to us of some vague impending disaster. How can we battle an enemy that won't show its face? How can we dissolve something that is so insidious and spread out that we can't locate its vehicle or its source? Again, it is consciousness that comes to the rescue. By focusing our awareness, we can ferret out the residues of fear within our being.

Past-life work is one of the most effective ways to "meet the source" of our fears; however, we can work energetically to remove these residues on a daily basis.

Here is an exercise that can be done anywhere, at any time, to decrystallize fear:

Place yourself in a meditative state by simply taking in a series of long breaths. Breathe in slowly and exhale slowly, drawing in light with each breath and exhaling thoughts and agitation from the body. Ask your body where it is holding fear. Simply ask the question, continue breathing deeply, and notice when you feel a sensation in some part of your body or hear the answer. Trust any sensation or feeling you get and go right to that place. Allow your consciousness to rest there and feel yourself moving deeply into the area, using all your perceptive capacities to connect with this area. How does the fear look or smell or feel? What colors do you see there? Then ask the body what color it needs to dissolve the fear. Accept the first color you see or hear or feel. Draw that color into the body with your breath and allow it to fill up completely the area with the stored fear. Keep drawing it into you until the body can receive not another drop. As you feel the color filling the area, simply be aware that it is absorbing and dissolving all the fear within you. You will know when it is enough. You will feel differently inside yourself—to some people it feels peaceful and lifting, to others it is a more active, ecstatic state. However you perceive it, when you feel the change, just open your eyes and go on. You can use this exercise in an anxious moment on the telephone or a fearful encounter. What is fascinating about this is that when you use the color to dissolve the fear, you actually dissolve the cause simultaneously. Each is linked to the other through magnetic attraction. *When we change the energy, we change the reality.*

As we become more conscious of our Higher Self and begin an inner merging with it, our body chemistry changes. Once the body has experienced higher vibrations and states of consciousness, such as bliss and rapture, it yearns to deepen such perceptions. If we are experiencing high frequencies, we are not "home" for the lower vibrations. They do not coexist. Our electromagnetic field simply does not attract them. This change of chemistry has a wonderful effect on the physical body. It very often spontaneously loses its desire for addivtive substances such as drugs and alcohol, which are so numbing to our being. The desire to eat meat falls away because on our inner level, we recognize the presence of negative energies and are repulsed. This happens because the emotional body is tightly bound with the solar plexus chakra in the area of the stomach, where ingestion of nourishment and digestion take place. At this center a new awareness allows the voice of the body to be clearly heard. This voice guides us: "I need more fresh greens and water," etc. We automatically direct ourselves toward a functional improvement of our system. We instinctively want a refinement of our energy system.

One of the most fascinating dichotomies related to honing energy is the use of consciousness-altering drugs. While there is no doubt that they change conscious perception, they do not effect the change by quickening our frequencies in a way that strengthens our energetic auric field. To the contrary, all of the recreational drugs actually put holes in the auric fields, which literally create a scarlike web that is to some degree permanent. Scars in the auric field disrupt the integrity of our energetic flow. This disruption is very noticeable in terms of our ability to concentrate or focus our attention enough to be successful at manifestation. The crystalline forms of hallucinogens do produce a speeding up, but it is of a kind that is erratic in nature, thus creating an imbalance between the head and the body. The traces of these chemical tricksters are visible in the iris of the eye for up to fifteen years. Meanwhile the slower vibrating substances such as marijuana leave a type of tarlike deposit on the nerve synapses in the brain, and we have not as yet found a way to remove them. Marijuana stays in the system for about forty-five days, which gives us the false impression that it is not addictive because the body will not begin to feel the loss for such a long time. It is a silent addiction, masquerading as a harmless way to relax—just as alcohol does.

All these drugs place us unaware and unprepared into the astral dimension where, indeed, we move as slow as molasses and at the cost of our own center. We are opened and un-protected from a myriad of negative, passive thought-forms and energies that stick themselves to our auric fields and go on disrupting our energy; determining that it will be impossible to soar to the heights of fast, clear perception and reality. The cost is simply too great to destroy the integrity of our auric fields in order to glimpse "God" only in a disassociated state, which we can neither integrate nor reproduce through our conscious intention.

We unwittingly turn away from consciously taking on the demands and tasks pursuant to our own self-arranged life plan. We cannot fulfill the challenge that we ourselves originated: the self-chosen birth in a certain form in a certain environment, our decision to learn certain lessons this lifetime, the possibility of giving ourselves and our gifts to the world.

If we use drugs, these challenges are not perveied. Through drugs, the polarity between these isolated mind-altering ex-periences and our everyday life is deepened. Our capacity for communication, our self-confidence and our strength to con-quer our laid-out tasks are dulled; we lock ourselves out of active participation in life. Drugs pull us continually further away from those people we love, and even worse, from our own Higher Self. (Consciousness-altering drugs don't teach us to gain entry to higher levels by our own free will and decision. They limit our capability to manifest our dreams and visions.)

We are *not* our addictions, our habits, our negative emotions or thought forms, and this is what saves us from such a bleak scenario. We only use them as crutches until we all, ultimately, begin to feel the unrelenting desire to touch something deep inside us that silently nudges a distant memory of who we really are.

By learning to stop a bit in our lives and connect with the Higher Self, we will find the guidance of true knowing, which does not plague us, but lovingly lifts us up so that we simply perceive ourselves differently.

It is fun to tune in and see what kinds of food we should eat today or what kind of exercise we should take or what part of the body we should focus on and clear. The body and the Higher Self are inexhaustible resources for exploring ourselves.

Meditation on the self allows us to experience wholeness in that childlike timeless way that lifts us up out of traumas and into a consciousness of wonderment and connection. The energetic shift brings a delicious vibrational "hum" that reverberates throughout all our subtle bodies.

NEW MEXICO— THE MAGIC OF LIGHT

10

We feel the oppression of time's heavy hand when we are experiencing something weighted down by pain or fear. But let a joyous moment caress us and it wisps away like the wind, leaving us reaching with all our might to recapture the tiniest scent, the tiniest glimpse.

Time exists within a sixty-second cycle only because our one-dimensional perspective tracks at that speed and holds it prisoner to a linear pulse. Any disruptive occurrence tears away the edges and multidimensionality rushes in to create an arc that totally alters time.

All of us have either experienced ourselves or heard stories of how time stalls into slow motion during accidents and life-threatening situations. I remember one time slipping off the shoulder of my horse at full gallop and falling on my back almost in front of the horse. I watched, in the space of my breath, the hoof rise into the air above me and slowly come down to meet my head. There was a soul-level conversation with myself about whether I wanted to live or not; all the reasons for and against, and a complete flashback of every major event in my life. As I made the choice to stay, the hoof landed a fraction of an inch to the side of my head and I heard myself exhale what I knew had been only one breath for that entire event!

Conversely, intuitive knowings come to us with a flash that is literally faster than the speed of light. Like illuminary fireflies, they flicker in and out of our awareness so tentatively that we only know of their presence after the fact—when we hear ourselves saying, "I knew that!"

It is this multidimensional quality that lends magic to reality. The more we ask inside for these knowings, the more they appear in our lives. Their value is not just the guidance they give us, but the way they affect and balance the emotional body, which then forgets its separateness and therefore its anger and pain. Intuition is a manifestation of synchronicity within our multidimensional hologram that extends itself to the universal whole. It acknowledges an attunement with "all that is." We can sense that space and we begin to feel harmonic. We all intrinsically recognize the rhythm of our own being and when we move with that pulse, we find peace.

The earth itself has many rhythms and is multidimensional in its reality. The interface between our rhythms and its own at any given spot is very much a part of the causality of what we can manifest in our lives. Whether we live in a city or a desert we must reach beneath the surface, cement or otherwise, and feel if the pulsating currents are beneficial to us. This attuning of our energy to that of the earth can influence the quality of our life—whether we experience happiness or dis-ease.

Everywhere in the world there are "power places." They all have their own unique energy and function to enhance and balance Mother Earth. They lie along certain meridians of this planet. Just as there are acupuncture points and meridians in our body, there is also a similar system for Earth.

There are indeed certain places whose specific energies can open a threshold of multidimensionality for us. New Mexico is one of those places. The quality of the astral energy there is such that all the experiences and events maintain their integrity in an ever-swirling pool of magnetic impulses. The time continuum holds no sway in the astral dimension, which simply deposits impressions that have a life of their own. The astral dimension exists in simultaneous space with the third dimension, separated by a thin membrane, which is easily pierced. Accessing the astral energies can be useful or harmful to us, depending on the clarity of our awareness. The timeless thought-forms can reconnect us to Earth's energies in a way that allows her to share her sacred knowings with us—in essence, we are the same.

Since the dawn of man's memory, the peoples of New Mexico have considered themselves caretakers of the earth, and the earth has rewarded them with a gift of magic. To show you what I mean, let me relate an experience that happened to me in the beginning of my sojourn in New Mexico. I had gone to a place in northern New Mexico to talk to an old Indian medicine man who was setting up a proposal to buy an old hot springs called Ojo Caliente. When I arrived he was busy with his business partner and suggested that I drive around for awhile.

I took the road north out of town toward an opening between the two mountains. As I approached, my heart suddenly began to pound and my eyes filled with tears. I became extremely alarmed at my body responding so intensely to something of which I was utterly unaware. Very softly yet quite distinctly, I began to be aware of a rhythmic beating sound like a thousand feet moving to the same drum. Voices chanting just at the edge of my hearing brought more tears down my cheeks. I drove farther up a wide canyon to a point where it closed itself into a bowl. Circling around, I pulled off the road to view the area. Without the interference of a moving car, I turned my attention to the vibrations. I could feel the presence of a crystalline substratum in the rock that always lends itself to magical qualities. I was just about to pull back onto the road when I saw three large black things pass by the corner of my left eye. My brain raced to interpret the data and the answer brought me to an instant standstill. Flying by me now in full range of vision were three pterodactyls! I watched them, mesmerized, until in a timeless flutter they disappeared.

I began to feel very strange and drove back to the medicine man's house. He came out to greet me and asked how I liked the area. In a hushed whisper I told him that no matter how strange it seemed, I knew the mountain was hollow and that there were people inside it, dancing. I closed my eyes and softly mentioned what I had seen in the canyon. He fixed his gaze intensely upon me and said the mythology of his people held that when this cycle of their human habitation of Earth began, they had emerged from a cave on the side of this very mountain. He said nothing about the pterodactyls, but again stared at me with a gaze that left no question that he knew exactly what I was talking about.

In the space of a half hour I had spanned two time-zone realities many thousands of years apart. I reviewed and reviewed with myself any triggers of association that might have focused my unconscious mind to either of those realities. Of course there is no answer as to why this happened to me, but there is a lasting effect: without thinking about it, my consciousness now has a frame of reference that can coalesce itself in any time-space frame.

Recently I was in frantic New York, riding along in a taxi. While absently looking at the faces in the street, I suddenly flashed on the pterodactyls and an entire, nonlinear yet comprehensive comparison of the energies was given me by my Higher Self. It was an amazingly thorough and descriptive study, complete with data about smell, texture, density. Not only was it fascinating, but it served to strengthen my capacity to be the observer, rather than allowing the kinetic energy to infuse itself with mine, and cause me to be overwhelmed by that singular reality.

Galisteo, the village where I live and the location of the Light Institute, lies in a large mountain basin, which had earlier been occupied by an inland sea. In nearby hills fossils of various forms of sea life have been discovered. The first time I came here I felt the special energy in this basin, which was undulant like the sea. Flowing in steady waves and swells back and forth here in this seven-thousand-foot-high desert town of New Mexico, I could feel this energy flowing like streams of water. Here in this place when one meditates on the ancient sea, one is tapping into a reality of millions of years ago that is still present—yet buried five thousand feet beneath the surface. It's a wonderful metaphor, and connecting with the salt water below the surface is my rejuvenation, since our blood is of almost the same solution. This aspect of electrical liquid energy is meaningful because a flowing of energy represents who we really are as energy beings.

Ridges of volcanic rock are elevated around Galisteo. They are constituted of black basalt and have an abundance of prehistoric petroglyphs. These drawings and symbols tell the history of people from the beginning of time up through the nineteenth century. Between two of these ridges energy vibrates and pulsates in a way most comparable to the echo of Tibetan gongs or overtones. The energy builds itself up between those two poles and creates a frequency that works like a vibrating energy spiral, or vortex.

Galisteo is, however, permeated by another power—the vibration of subterranean crystals. Galisteo portrays an energy spiral that is capable of bursting forth out of time and space. Galisteo is, according to meteorological maps, one of the places, maybe even *the* place, with the least amount of atmospheric pressure! It works like a window to the sky. I learned about this at the time I was in the middle of my acupuncture work, and thus perceived it as a synergistic proof, part of the cosmic game, that I should use the esoteric "windows to the sky" acupuncture points in a place that itself is literally a window to the sky due to its low atmospheric pressure!

Since I have lived here I have had a lot of encounters with the special energies. The sacred energies of the Indians and animals speak to me and have taught me whole systems of symbology so that I can better comprehend the signs and signals coming from the material world. For centuries medicine men have gathered in the Galisteo Basin to make contact through this window to realities beyond the third dimension. The extraterrestrial galactic energies use this special frequency for contact with Earth. The petroglyphs around Galisteo are full of depictions of space beings with whom the "Grandfathers" have conversed. Such energy currents can penetrate barriers and manifest the formless on this earth.

Though I had seen the strange lights gyrating in the sky many times, and even had my car stop in the presence of a huge spaceship in the night, I had treated it all as something natural, yet distant. Inevitably the morning after the sightings there would be Air Force helicopters out surveying the exact area where I had seen them. I knew that they were tracking them as well. One day my daughter Megan, then age eleven, and I paused in the car before a left turn near the church, on the way to get the mail. I had paused briefly to see if the road was clear to turn. It was a radiantly clear, sunny day, high noon, and we weren't in any hurry. Suddenly, from behind the church appeared three small silvery spaceships. They flew about six feet in front of us, so low that the treetops on the hill above the church were visible behind them. I first thought these were from the Air Force base located about seventy miles away in Albuquerque. I wondered why they were flying around here so low! They flew at a level of those few feet above ground right between us and the church, directly across the basin of Galisteo and toward a hill formation we call the wave. Two of these unusual airships flew directly into the wave! They simply disappeared into the rocks of the hills. As my eyes followed the third ship, I visually recorded how it hovered a moment in the air and then disappeared into nothing before my eyes. Megan and I sat a moment, transfixed. Then she said, "Mom, they didn't have any wings." They really had resembled little blimps. I replied that they had made not a single sound in the silent Galisteo basin. The silvery surface resembled quicksilver. We simultaneously realized that we'd just seen three UFO's.

What can you do with such an experience? We both had perceived the same thing in broad daylight. We couldn't rationalize it away. The reality of it wouldn't allow us to retreat into one of the rhetorical explanations used for UFO's, e.g., an unusual weather formation or a secret airplane prototype. What are UFO's? I think that some of the UFO's may be man-made prototypes. Many, however, really are what they are perceived to be: bridges of communication between galactic societies and ourselves here on Earth. They are coming from civilizations of another sector in space and they are willing to establish contact with us. Such UFO's are actually light-body portrayals, rather than dense material like steel. Their medium of contact is through these forms constructed of light.

Galisteo and its unusual energy field contribute decisively toward the transformations of our various bodies, right down to the molecular structure of our cells. The higher vibrations of the region pull one into an evolving spiral, winding upward into new dimensions of consciousness. These vibrations also enable us to participate in that occurrence hiding behind the phenomena of the UFO's. UFO's move at a speed faster than light, back and forth between our space-time and negative space-time. They enter the formless beingness and return in definite forms of material existence. There have been stories aplenty about UFO's in all times and cultures, and we all too hastily classify them as "mythology."

New Mexico is a land of radiant lights. Here in the northern part at 7,000 feet elevation or higher, the sky is often deep blue under a glistening sun. The light takes on a special quality. It is almost like living in the shine of a rainbow. Almost all the rainbows here are double, and their color energies are immediately sucked up by the earth. I once stood in the reflection of a rainbow seeing and feeling its color frequencies cover my body.

We are composed of light. The development of higher consciousness must include the development of our own light body, with which we can penetrate walls, be at two places simultaneously, appear and again disappear. What still must be learned is that our body vibration *can* pulsate so fast that we consciously perceive and use our light body. Modern science has also realized that matter is a crystallized form of energy and is portrayed in different energy forms microscopically in a rhythmic light-dance of molecules and atoms. There is a vivid synchronicity between the light-filled macrocosm and the likewise light-filled microcosm.

We have heard about ecstatic saints who could appear simultaneously in two places. Dematerialization and material-ization (one thinks of the miracle saint, Sai Baba) have frequently been witnessed by many people—and they will continue to be. What about the appearances of mystical angels, or of the many Madonna appearances? Jesus also used his light body to appear before his disciples. He knew of the cosmic laws and how one could manifest the ethereal self. These are all processes in which light bodies appear and the patterns of light energy may be used. What is so magnificent is that, at last, our consciousness is evolving to a level at which we can utilize these "cosmic laws" on the physical as well as the spiritual levels. These miracles are simply laws of energy that we can begin to comprehend by extending our consciousness into these higher realms. We can learn to use these principles in very practical ways. If we materialize healing substances, we can dematerialize destructive ones, such as radiation or pollution.

Despite its outwardly desolate appearance and far-removed location in the high desert of New Mexico, Galisteo is unusual in its energetic effects. Of course, one can gain experience with one's light body in other places in the world. But here it becomes much easier to release oneself from old limitations, prejudices and hindrances, and open oneself up to the new. The energy of the light and the force of the energies are intense. At the Light Institute I notice how clients begin automatically to extend their auric fields. They are quickened by the energy and their auras noticeably lighten.

A few years ago there was an important solar eclipse. Many scientists traveled to Kenya in order to glimpse it better. At that time, I was also in Kenya, with my daughter Teo. We were *exactly* in the area of total eclipse. I can remember how everything around us took on an odd yellow coloring, the air stood almost still, no birds sang, no insects buzzed, everything was suddenly deadly still. I felt as though my eyes could see through everything—like X-ray vision. Because of the unusual energy of the eclipse I could see through the tree leaves, through blades of grass. I perceived everything around me with penetrating clarity. This kind of experience happens a lot in Galisteo even when there isn't an eclipse, because of its multidimensional energies. Here one can see the prana in the air and can see through material structures, even though the ground is very compact and the volcanic rock formations are dense, like the cacti. The crystalline substructure of Galisteo affects our perspective when we allow ourselves to experience magic in our natural world; we create it in our daily lives as well, by being able to look at events and ourselves in a more realistic way.

The close observance of a leaf—simultaneously perceiving all inner and outer layers, and seeing right through it—is very comparable to the process of death. The death experience is a passage through hitherto impenetrable barriers. Our consciousness experiences the transparency of the different dimensions of life and realities, seeing through them all. Consciousness itself is a bit like fog, like fine glistening prana, a collection of the finest particles vibrating in harmony. Death is a channel or bridge for this consciousness, leading to higher and greater realities.

DEATH IS AN ILLUSION, TOO

11

Perhaps the cruelest of cosmic jokes is the illusion that death lurks around every corner, perpetually measuring out our life as if each experience or feeling must be paid in kind to an insatiable force. We attempt to cheat it of its deadly staples by fending off the living—by becoming numb. As if we could save ourselves by pretending not to move. Within the deepest essence of our knowing is the unmanifest, disembodied self. Yet it is the body that holds sway in the realms of outer consciousness, that whispers its fears of certain ending we so ardently strive to delay. The energy we expend by holding the resistance to that ending is the energy drained from our life.

The body remembers the dying it has experienced over and over again. Since it is rarely available to witness the hologram of our multidimensionality that death initiates, it has no consoling cognizance of perpetuity; it is simply caught in the agony of the dying. This is why so few of us actually struggle with the concept of the finality of death itself. Yet our intellect cannot save us from the nightmare of our worst fears about how death will overcome us with its unrelenting waves of dying: Will it be disease? Will it be painful? Will it be long?

Death becomes the focus of our fear only because the trepidation of pain to the body is unbearable to the mind. Thus it retreats to the more abstract ponderings of finality versus eternity. Indeed the mind is haunted by the unconscious repertoire of the body in terms of its physical demise. We have rarely treated the body with grace in the dying process. The violence and disregard are indelibly laid down in the cells and passed on in the same way as the genetic coding is passed on.

We are afraid of pain. Our own fear creates the energy

blocks that ultimately leave the body so contorted that it manifests the disease and the new pain, which we were trying so hard to avoid. Each time we reveal a death we have experienced in a past life, we allow the body to release the energy it has stored and its memory of the pain in dying. Very often, for example, just the recalling of a life will trigger the body's memory of its physical experiences in that life. The person will begin to feel sensations, and as we bring the awareness to these sensations, the body releases the stored energy. This has a wonderfully beneficial effect on the present body.

In working with past lives, one personally experiences having died many times! As soon as we can release the impressions of the death experience, which are imbedded in the cells, we feel as if we've been freed. We are free of the fear that life stops with death. We are our own witness to consciously surviving outside the body on different levels.

Through this work with past lives we can wash out the emotional as well as the physical residues. We can release ourselves from the unconscious fear of certain situations that resulted in former deaths and that our current body remembers. For example, perhaps in this life a person suffers from migraines and worries about their cause. The person may recall the source of the head troubles in former lives, such as a lifetime when the head was smashed, another time falling head-first into a rock embankment, and another time of brain hemorrhage. As the body tells the tale, the imbedded impressions are discharged from the molecular structure and the head will feel as if it has been freed. Sessions often reveal that clients with heart trouble, digestive problems or other chronic ailments had, in past lives, suffered from fatal injury or disease to that body part ailing them now. In many cases the clarification of difficulties creates lasting physical relief, or even healing.

Everyone feels afraid or uneasy about death—consciously or unconsciously. One must keep in mind that in our country we don't face death experientially. We shove it aside in hospitals and cover it up under white sheets. We transfer the responsibility to the medical personnel and maintain a safe distance as "innocent" bystanders. We don't want to take part in the confrontation with death because our fear of it is so deep-seated.

Such experiences corroborate the realization that we do not fear the unknown; we fear negative experiences we've already had, even if they are now subconsciously buried. How often do we automatically feel an impulse to protect our stomach or head or back or some other body part without a logical reason? The Middle Ages are the best, if the most gruesome, example of how we have let ourselves become fascinated by the materiality of our physical bodies, when we have entangled ourselves in bodily fear. As we manifested into the physical level from the spiritual level of light bodies, and multidimensional being, we were overpowered by the attempt to define ourselves only through matter. We quartered, beat, tortured and maimed our fellow human beings. We lost ourselves in sexual exploitation. All that because we wanted to "understand" something that in truth wasn't material, but rather belonged to another level of our consciousness. We tried to tear away the veil between our unconsciousness and ourselves by peeling away the physical body. We had a feeling that our life didn't end with our deaths. We had a certain hunch that life energy expressed itself in sexuality. We tried to squeeze and press something out of the body, which in actuality issues forth from purely spiritual essence. We were looking for the soul. We could sense its existence, but it eluded us.

This recognition of power in the body and its abuse in the Middle Ages is still reverberating in us today. This is one reason for so much physical suffering, so much karma on the physical level, and its resulting deep-seated, pervasive fear. Our body throws such memories back at us like an echo. It makes itself felt during sessions, drawing attention to the organs and places that are energetically charged from experiences in earlier lives. Then the body's own energy system ensures that such memories are flushed out and discharged. This usually results in a feeling of liberation, weightlessness and luminous life force. The body is at last released from the addiction to repeating old patterns of experience.

If we release ourselves of a certain energy, we create simultaneously a new energy, since we operate in energy fields and between energy poles. If people let go of their fear of death, setting their old associations free, a new energy flows into them automatically. This new energy expands the consciousness. One is elevated to previously unknown levels of understanding and realizes that birth and death are nothing more than pulse beats in the universal life of that individual consciousness, that they mean nothing more than passage from one level to another.

Here is a beautiful past-life sequence in which someone was using the physical body to teach these principles to others.

▶ "I see myself lying in a sarcophagus. It's Egypt. There's a ceremony as if I'm a great person, like a king. The impression is, I'm more of a priest. I'm young—

about twenty-seven. There's a great celebration. I am an important person.

"What I get is that my function as a priest was to prepare myself for the death, and to communicate after death. The king has a great fear of dying. Because of his preoccupation with death, he is unable to function effectively as a king. He does not pay much attention to the economy, the security of the people. This is rendering him ineffective, leaving the affairs of the kingdom unattended.

"As I cross over to the other side, I begin to communicate to the king through dreams and meditations. I tell him that death is no different from life—it is just on the other side of the veil. We are still the same. No different. I encouraged him to experience and value his life on earth—that it was no less important than what happened after death. That it was all connected—all important.

"As a result of my communications, the king dropped his fear of and preoccupation with death and began to truly live his life as king. My job done, I then moved on into the light."

What a magnificent gift to mankind—to unlock the mysteries of death and lead us boldly into the hereafter! Within that vast unmanifest tide of consciousness we can perceive life from a totally different perspective. What we experience after releasing the body depends very much on our emotional state at death. When we are at peace, the consciousness moves into octaves of rapture and light; feeling itself enveloped within the great all. If death is accompanied by great emotional turmoil, the spirit often becomes caught in the astral levels. Here, it forgets itself and is often literally bound to the very spot on the physical plane where the emotional body was imprinted by the trauma. This is the case with co-called ghosts and spirits. We can release these beings by making contact with them and showing them the direction of the light, so that they can move again into the pulsating spiral of evolution.

My daughter Teo, age six at the time, and I shared a powerful example of this once as we left our ancient adobe house that sits upon an old Indian ruin. As we entered the car, she suddenly said, "Mother, someone is in the car." I looked around and replied, "I don't see anyone." She had meant a spirit. I suggested that she ask this spirit who it was and what it was doing here. She said that it was very afraid and didn't know where to go or what to do. She related the entire story of an Indian spirit whom I recognized as having appeared in and around our house before. I told her that she should imagine the whole car full of light, and then advise the ghost to recall all his relatives who had departed the earth before him. When he felt them, he could simply step into the light and stretch out his arms to be received by them.

Teo did that, and we both distinctly felt the energetic shift as the spirit was freed. He had become entrapped in the astral dimension because of a violent death. As soon as my daughter let in the universal white light, he was freed of his fear and very definitely reached those beings he loved.

Shock, anger and vengeance always bind the spirit. Many times we may even reincarnate into a new environment, yet leave a part of our spirit imprisoned in a former life. Often we make vows and contracts to pursue our revenge that per- petrate destructive relationships because we refuse to release those with whom we are angry. It is not unheard of actually to inflict that punishment on those still in body even after we ourselves have died. This is a great misuse of personal will. Here is such a case history.

Marie was in the process of divorcing the man who was her husband in this lifetime. Experiencing this scenario helped her to release not only him but also the feelings of anger and vengeance that had always been a part of their relationship.

▶ "Our relationship was cold. We were married because our families set us up. We didn't love each other. He knew I was going to kill him someday. I hated him because we had a son and he killed him. The son had something wrong with one of his legs. My husband believed it wasn't right to let him live . . . that any son of his had to be strong and powerful. He waited until the baby was one year old. He took him when I wasn't there. He had no compassion. He was just cold about it. I feel such a loss, heavy grieving for my baby. Years later I had become hard and cold. I couldn't have any more babies. He didn't like children. We no longer had a sexual relationship. He was seen around with another woman and everybody knew. He didn't care if everyone knew. They would get drunk together. I hated both of them for that. They were disgusting. I got older. I had so much hate. My body changed; my right hip was higher, I started bending sideways and my back was crooked. I'm feeling sad and victimized. My face is twisted and my left jaw is real tight. I'm no longer living with my husband. I'm living in a shack. I went crazy. I just wandered around gathering herbs and being pitiful. I talked to myself in public. I was still thinking about my child. I talk to my child; I pretend that he is still there with me. I had my son and nobody knew. I laughed about that. I was better off than they knew.

"I was walking up on a steep place and I fell on the rocks. I died a lonely, old, miserable woman. I fell on the rocks and hit my forehead. I lay there a long time before anybody found me. At death I felt anger at myself for not having done anything. Being so angry, I took it all inside and hurt myself because I never expressed it. I felt stupid. I feel that I stayed around to watch to see what would happen. I hoped my husband would have an awful death. I made him have an accident on a horse. I spooked the horse, and he reared into the air; my husband fell to the ground, and the horse's hooves trampled him. I saw his girl friend come up the hill where my husband's lifeless body lay, crying and throwing herself on his body. I had so much hatred for the two of them. After seeing the pain I had created, my revenge was not so sweet. I felt sorry and guilty. I learned that I shouldn't be so vengeful, that the consequences are too severe. Finally my soul was released from that lifetime's karmic ties. I could now go into the light.

"A kind of celebration took place, where the people told stories about the loved one. They sang and danced on their behalf. The people sensed that the soul of the deceased had been freed, and acknowledged that freedom with great reverence. At first I was astounded at how easily they let go of their dead, be it a spouse, a parent or a child. They gave their pain and grief free rein, but then they let the dead go! It was completely new for me to experience how sorrow, letting go of the deceased, and celebrating their liberation could be intertwined."

The most important teaching I have received in my life occurred during my work with the Peace Corps in Salvador. At that time I lived in a small village named Dulce Nombre de Maria. Many generations have remained settled there, most of them never having been outside their village or having traveled to another town. I worked in the health services and saw many people die. A certain form of tetanus and other illnesses were rather widespread. (In Paraguay, where I later worked, almost a third of all newborns died.) Children and babies died regularly, a few in my own arms. I held God responsible, and demanded to know the Justice of it. Why should these innocent little beings die, often in such painful ways? I observed closely how the relatives of these children and babies behaved before and after their deaths.

Here in these villages death is a component of the everyday ups and downs of life. If someone died, there were naturally spiritual accommodations for the mourning, to lament the loss. Everyone took some part in expressing the suffering that the deceased, the family and the community had incurred.

We can learn to let the deceased go and recognize that, in reality, no one is ever lost to us. On the spiritual level, we meet one another again and again! If one becomes freed of the body, the connection on the spiritual level is often made easier. There is no separation. Whether we are in body or not, the mechanism for spiritual communication is available. It is the nature of consciousness.

Those children dying in my arms provided me with the fundamental motivation to research the multidimensional self. I had to find out why our lives played out this way and according to which principles our lives functioned. I wanted to learn more about why the death of one person could guide the spiritual development of another. I still needed to understand that the point in time, conditions and method of death went according to one's own plan, made at the spiritual level. I learned that the deceased wasn't the victim of some external conditions. I felt that I needed to test my own conception and projection about death—above all, I needed to test my viewpoint that the deceased were "victims." I discovered that we could access the Akashic Records of people, and gain insight into how and why they selected their death experiences to clear karma and further their soul's growth. With this new knowledge, my old conceptions about death appeared inadequate and I discarded them.

How death completes the lesson of a lifetime is of great significance to each individual. We ourselves choose our method of death. We choose how we depart this earthly level, and that choice is made at the spiritual level. Cancer, for example, has been recognized as a form of suicide for a long time. (It is broadly known that to a large extent, the origin and development of cancer depend upon so-called psychosomatic factors. We allow our body to become disintegrated inside. The power in gaining this perspective is that at any point when we consciously acknowledge and heal the source of our disconnectedness to life, we can erase the disease.)

The reasons for and the circumstances of our deaths give the soul a chance to accumulate certain experiences and to release certain deposits in our emotional bodies. We must learn to honor those choices in each other.

Let's examine the various killer diseases for a moment. In the past they were agents of external attack, such as infections,

plague, heart–circulatory-system illnesses, which signaled a
new era in which the vehicle of the body simply could not
withstand the stress placed on it. Now cancer, AIDS and
possibly radiation sickness move to the forefront. These diseases
are all symbolic of inner dissolution on the more subtle levels
of the body, i.e., the cells. The illness strikes the individual
cells throughout the body. We must establish that the departure from this earth no longer results from clearly definable
single causes, but rather that the cohesive glue of our physical
integrity weakens. Cancer, AIDS and radioactivity directly affect the molecular structure of the cells, creating a change in
the energy flows therein.

We aren't here just to die of AIDS or to see half of mankind
degenerate away. We must ask ourselves why we have chosen
AIDS. We are all collectively responsible for everything that
occurs on the Earth. If we approach AIDS from this standpoint,
instead of shoving it away, we will find practical assistance
very quickly. Spiritually, these diseases are a gift to us. Their
purpose is to teach us about death, about how to be fully
present in life. As this trio of AIDS, cancer and radiation
sickness spreads out into all octaves of human experience, we
will become intimate with death and awaken to compassion.
If we face it head on, we will recognize the spiritual force
there. By participating, we will learn, and in learning, we will
become the masters who manifest on a co-creator level.

The suffering, natural catastrophes, wars and fears that we
currently experience, correspond to the prophecies of olden
days. Despite or because of all their negativity, they create a
very necessary clarification and cleansing. They stimulate our
consciousness, challenging it to find sources and solutions to
these problems in such a way that a positive future is developed.

These negative forces are like "propellants," thrusting us forward into a dimension of light. Our whole body becomes involved in the transformation process, which occurs throughout the world with ever-increasing speed. Our planet becomes seized by a brand-new energy. We encounter new forms of illness that mirror ourselves. They lead us toward finding our center again. From this center we can radiate new strength through the body, establishing harmony and balance.

When loved ones die, it is crucially important to release them. Their transmutation from this plane onto the next is a graduation. We must get above our own lessons here so that we too can graduate. Their emotional body needs to be imprinted with the experience that we have released our karmic bond with them, lest it continue the same emotional issues, perhaps projecting them on whoever else is around. It is important to visualize that they go into the light and disappear, so that the emotional body realizes that a change has occurred. We may also discover where in our body thoughts or feelings relating to them have become lodged, and release them as well. We must celebrate their newly won freedom with heartfelt thanks and grace. When we can let the dead go in this way, the essence of their being remains united with us. The highest vibrations of unity between the dead and our self become a new, conscious component of our beingness.

My mother died shortly before I went to Africa in 1985. I'll never forget the all-enveloping light that permeated my home as her exuberant spirit came to give me the news. For weeks afterward I felt as if she were seeding me all her wisdom and grace and love. I spoke a long eulogy for her at her funeral and I sensed her attention—yet from a great distance. She is now so near, it is as though she moves my hair. I feel no separation between us. If I missed her earlier, I had to call her by phone. I no longer feel the separation. This experience profoundly changed me.

When we take into our being the essence of a parent, we become the parent. This is an important understanding in the case of early separation in childhood. The child, then, has accrued the power to fulfill the role and actually does not need the parent to mirror it, but must assimilate it within the self. I became the "mother" in a new way!

Our preparatory sessions, in which we work with the emotional body, often involve releasing others on an inner level—setting them free. Just as the people in my Latin American village didn't look back after mourning and celebrating, we too should no longer mentally or emotionally hold on to our dead. If we constantly look back and hold on through our grief or other emotions, then we literally entrap them. It is as though they become "possessed" by us because of our clinging. This interferes with their development. If the emotional bond, whether love or hate, between the living and those who have passed on is very strong, the ties are sometimes very binding, even to the point of making vows and contracts on both conscious and unconscious levels to continue the relationship forever. These promises and thought-forms manifest themselves in other incarnations, which we chose in order to fulfill them.

Entrapment occurs very often with abortion. Women feel guilt after having an abortion and develop pelvic difficulties and illnesses. Abdominal infections occur very frequently. In the last twenty years I've only encountered two or three women who were open and clear in the pelvic region. Behind the physical disorders are usually lack of communication, misuse of sexual energies, and abortions. All this contributes to an energy blockage in the abdomen, resulting in tension and illness.

Women who have abortions are often caught up in the "yes-no conflict" between wanting and not wanting the child. The body always wants a child. To conceive and bring forth new life is its innermost biological drive. This yes-no conflict is very confusing for women. Their guilt often leads them— even if unconsciously—to hold the spirit in bondage. The woman may think that she can't raise a child now, but perhaps in five years. She then does not fully release the child to incarnate through someone else. Instead they both stay locked together. Perhaps it is part of his life plan to be borne by this mother into the world. If his birth is postponed by the mother until later, the mother-child relationship may be patterned from the beginning by the mother's guilt and the child's un-certainty about incarnating. Of course neither is the victim of the other, but by their emotional ties, they chose to be in-tertwined.

We can dissolve this dilemma by using spiritual awareness to allow communication between the two so that entrapment does not occur. The pregnant woman can converse with the entity so that both are clear about the choices they make. What a difference this creates in women's bodies!

There are so many souls who want to incarnate on this planet at this time. Because their purpose is to help us through this phase of enlightenment, they have less karma and more power at their disposal. Recognizing this will eventually rev-olutionize the nature of relationships. They know that they will not be in adult bodies for the quickening into enlight-enment. We will have to treat them differently than we have treated children in the past. Imagine master beings directing transformational energies from the bodies of babes!

We are in a stage of development on Earth in which every soul can achieve mastery over his own growth. Every soul now embodied on Earth did so by his own choice. He also chose the conditions of his existence. Every soul carries within itself the highest ability to understand self and to decide what is best. This applies to spiritual development as well as to interrelation with others and with the environment.

At the Light Institute we have worked a lot with people who confront death and want to prepare themselves for it. It is always a momentous revelation when they discover that death is merely a passage to greater, higher and lighter dimensions. They ascertain that death never leads to isolation, darkness or to a hopeless dead end. They realize that when one is capable of consciously manifesting conception and birth, one can also master death. They begin to open up a timeless energy field in themselves.

This energy field facilitates their future passage over the threshold of death into other dimensions of consciousness. During the sessions they have wonderful light experiences, karmic solutions, answers to questions about life and about lessons in consciousness. They can cross through this tunnel to a bodiless reality fully prepared and without fear. Since they recontract to so many different former lives and deaths during the sessions, they learn not to reject death, suppress it or become frightened by it, but instead to approach it with ease. They orient themselves toward mastering the passage of death as an "art" with awareness and grace. They work at maintaining continual contact with their Higher Self. They dissolve their illusion of time, separating themselves from the false security of linear thought. They learn to understand the whirlpool of creativity within them. This energy vortex contains the formless knowing. They let themselves be seized by this energy vortex and radiate from within.

At the Light Institute people begin to discover the process of death as an integral part of their life and consciously seek to know their karma, then their tasks in life. Because of discoveries like these, more and more people will decide to die at home and not in the hospital. Family and friends will be given a chance to understand the death process. Having a loved one die in your arms changes not only the deceased but, above all, the survivors. If we embrace death as a natural process, understanding it as a natural ending, then we too can achieve this perfect mastery of life and death. The end of the process, the translation of matter back into energy, has its own meaning as well. We must allow the circle to come to a close. If it is closed we can let it go.

After people have cleared the source of their karma in this life, it often happens that they turn toward the clarification of themes concerning dimensions other than our earthly life. They realize that life isn't ended or closed with death, but that it proceeds in other areas of consciousness. Perhaps they recall themselves in an extraterrestrial body. I have worked with someone who remarked in session that he hated ears. He discovered that he had belonged to a group of extraterrestrials who didn't have ears. He considered ears grotesque and couldn't adjust to them because he had been accustomed to direct, nonverbal communication not requiring ears!

Many people wonder if they will be born again. People who can work out current problems by seeing their sources in earlier lives can certainly glance forward into their own future development. Many of our clients have received messages from their Higher Selves that indicate that they needn't return to a human body on the earthly level. If one has achieved a balance of energies within the self, if one has learned the powers of manifestation on this earth and their correct applications, it isn't necessary to take on a new body in the compact material sphere of this planet in assumption of a new role. Instead, such souls go into other development arenas, they experience dimensions of consciousness in other universes and galaxies. It may be levels of light or sound. From our limited standpoint, these regions would be identified perhaps as different arenas of enlightenment.

How can one confront and prepare for death without sessions, past-life therapy or other external aids? How can one deal with fear of death? Being in the loving place of the Higher Self is a profound way to relieve the anxiety. It brings us back into contact with the unmanifest soul.

I have helped many people literally to practice moving into the passageway of light so that they have a sense of orientation. People who are dying almost always tune in to their Higher Self naturally. From my observation, fear sets in during the beginning phase before death, the initial arena of bodily reaction. As one comes very near the passage, one becomes peaceful. I recommend that one practice contacting the Higher Self. Just as the Higher Self takes form, focus on that form. Draw it into your body so that the frame of reference is holographic in your consciousness.

I must share with you a letter that is one of the most touching examples of working with the death experience. It comes from a young German girl I met at one of my seminars in Germany. I'll never forget her or this letter.

▲

Dear Chris,

I have to write to you—and I want to do it the
way I feel it, open and right away
just in accordance to this strong connection I felt to
You—
being known to You since a long, long time.
To make You remember me—I have been at Your
seminar at Frankfurt in August 1987. In a break, outside, my question to You was, if I could overdo it
with my mother, who was seriously ill from cancer.
Your answer was meeting my third eye right away.
"No, you cannot overdo it. It's your karma.
You have to be gracefully
and full of light."
And you laughed into my heart and my eyes.

(So, I take the typewriter now, because it is easier
for you to read.)

So I felt totally free to give all my love to my
mother.
I gave everything I could give
in this last period of her lifetime.
Somehow I arranged that she could spend the last
month at home, which was what she wanted most—
to stay with her family, but she didn't dare to say.
My heart was totally open
and so I got some help from the sky, a connection,
which is staying with me since this time.
I felt in the end, that it wasn't me doing all this,

with this little sleep, exhausting job, something was
flowing through me—I just had to open up and follow my intuition, nothing else.
My mother and I developed a kind of contact
which was similar to the first blossoming in spring-
time.
I gave her body, who was just skin and bones, naked,
without any hair left, without control—TOUCH and
ANYTHING my hands felt,
I hold her in my arms and stayed and stayed.
And sometimes she suddenly opened her eyes,
her dark blue eyes
like crystal balls
looking into this narrow world
which was now only this calm bedroom of my parents
with surprise and joy like a child
with these sparkling blue eyes
I could have not even imagined before
and I knew for sure
this is my mother—it is her.

Listening to her breath, as usual in the night,
it turned out to be her last night in this lifetime,
her breath changed totally
just from no to yes.
From this heavy, anxious "no, no" and all her sorrows,
her worrying and her feelings of guilt and this incredible
resistance, which was with her for nearly all her life,
she turned to yes.

    Yes.

A sigh from relief
on every exhalation
so deep, so soft, so warm, that my tears were running
with her breath.
My brother arrived in time
and my Dad, who was overwhelmed by the sadness
of his heart (he is a quiet and shy person) got this
chance,
to tell my mother all the things, made for HER EARS,
these words, she was waiting for all her marriage life.
I told him, that she couldn't speak anymore,
but that she was still listening to us—
and so he talked quietly to her
and wished her a good travel
and said all this with his silent love and respect.
My heart wanted to break at this point.
My father holding her hand,
my brother intuitively touching her heart
and me sitting at her feet holding them tight
she gently looked into my father's eyes
the last light of her blue eyes upon him
and slowly, slowly she started her way back home
through the light into the light.
With the light of the first morning sun on her face
the day, which was the last and beautifullest day of
summer in Germany this year
and this big, big silence around her, she left.
With a carpet of quiet tears
and love from our three hearts
she left so easily, so free,
and her face was showing that. . . .
The air was filled with this white glowing light I know

so well now.
Any separation
I ever felt was gone, in these hours with my family.

Later on I washed her body for the last time
put on her beloved clothes
the shoes in which she felt best
and gave her the smell of her perfume,
which was her favorite . . .
No, there was nothing left to say or to do.
I thanked her a last time for this body and this
chance she gave to me—
nothing left but
    LOVE.

▼

A GLIMPSE INTO THE FUTURE

12

What lies before us, what effects are produced by working with consciousness?

The major transformation of reality comes from the realization of our intrinsic participation in all that is. When we experience the power of synergistic relationships in which energy (thoughts, emotions, patterned pulsations) translates into matter—crystallizations that create the actual experience of form (health, disease, catastrophe, ecstasy)—the true hologram begins to emerge.

The limited concepts of past, present and future dissolve the illusion of their fixed position on the cosmic wheel of life. Nothing that exists either as energy or matter is stationary because energy and matter themselves are eternally interlocked in the blinking beat of divine consciousness.

It is only the fearful resistance to this motion that causes evolution to fall back upon itself and collapse into "history."

We are able to comprehend that the future is positioned upon the building blocks of yesterday. What we don't recognize is that there are unmanifest components that parallel and interweave with the manifest that provide unlimited material for tracing the horizon. This base expands itself into ever widening arcs of probable realities that correspond spacially to each other and are the infrastructure of what we call the "future."

Thought forms adhere to certain strata of crystallized matter according to the emotional grouping. Negative imprints associate together in the principle of "like attracts like" and never intermix with the faster frequencies of positive emotions. As they clump together, they literally form new but similar realities out of their very substance! It is a startling revelation to discover that thought forms are not inanimate, lifeless particles of matter, but are actually organic, life-force energies with all the procreative charge necessary to produce themselves in kind. If we become the witness of this phenomenon, we can participate in the future by consciously organizing the energetic components of our thoughts so that the new groupings are of a powerful, positive nature due to the quality of the source material.

This challenges us to free the crystallizations that stifle our potential. By simply shifting the weight of our attention to embrace the unmanifest space, we can open the consciousness to the new cosmic now. The magnificent power of meditation lies in its effect on our time-space reference. It dissolves time, which facilitates the experience of endless space and light: the eternal now. Bathed in these frequencies, we can easily be totally present because nothing is missing—thus worry does not exist.

Whenever we attempt to envision what is waiting to form we bind ourselves to the limited mind because we are only conscious of what we already recognize, not the contents of our multidimensional hologram. If we fill the space with light, the subsequent creation will always be of a high vibration. As we become aware of our negative programming and our anxiety about the future, we can dissolve the very energy that is creating what we most fear.

Exploring our multidimensionality awakens us to the consciousness of choice. Each and every experience taken on by the soul is intrinsically one of choice. As we revisit past lives we review those choices, freeing ourselves from the grips of repetition and judgment favored by our slow, unrefined emotional bodies.

We can learn to use choice as a tool for growth. The more conscious we become that we are indeed choosing our reality—and that the will of our intention brings it into form—

the more we can begin to train that will to bring about choices that lift us into higher and higher octaves of ecstatic experience. Thus, we see the true significance of our experience. If we look upon each one as a gift, we soon escape the trap of the victim.

It is a wonderful exercise to occupy the mind in these terms: actively seeking out the gift in any seemingly negative experience. The conscious awareness that we can use our experiences for our higher good frees us from the fixation of powerlessness, which denies on such a profound level our conscious participation in the future.

Experience from past lives also can be used to strengthen our repertoire of abilities. After returning from a two-year tour in Bolivia, my son Britt, who was about eleven, found himself swallowed up in the American system of math. He felt blocked in the subject and suffered very poor grades. As he experienced his first past-life recall in a session he discovered that he had been an architect for one of the large Egyptian pyramids. Britt hadn't ever read or heard anything about the pyramids. I was especially surprised that he described them in detail. After the session we didn't discuss or analyze the images and experiences he recalled. We took it simply as an interesting story out of an earlier life. After two months Britt went from being below average to an excellent math student. In the next semester he was even invited into the accelerated math course. Since then he has remained at the gifted level and intends to study quantum physics.

This sudden change was caused by a basic transformation of his own self-perception. The revelation that he possessed exceptional abilities totally released him from the grip of his self-doubt and pain. His glimpse into another incarnational reality dissolved his current blockage by giving him a positive frame of reference under which he could establish a base as a successful problem solver.

We are in the midst of a consciousness revolution. The approaching frontier is in the research of the higher self. If we strengthen those experiences and amplify those situations that lead us to conscious mastery of life, our everyday consciousness also becomes oriented toward possibilities of telepathy, self-healing and the application of many unknown and even unimagined creative powers. The exploration of our consciousness is the most breathtaking adventure of all. If as individuals we take on the mastery of life and begin to seek the valuable contributions we were born to make, our consciousness will no longer be busied with fear. I see us on the threshold of manifesting a new perfection. Our emotional body can be focused to experience and radiate ecstasy, rapture and bliss. Each flicker of enlightenment within the individual contributes to a deep-reaching transformation of the whole. If we become one with our higher self, worlds of creative effects open up. We can cognate the limitless wisdom of the pulsating, far-reaching universe, both on levels of manifestation beyond human understanding, and we can also, through our collective conscious latticework, pluck the knowings of our greatest sages throughout human history. The attainment of any one individual is available to us because of the commonality of the source, i.e., universal consciousness. Contemplation of great beings attunes the mind frequencies to these bands of frequencies used by them. We forget our self-consciousness in the presence of all-encompassing wonderment. We must perceive ourselves as naturally inheriting the attributes of the great models of mankind. As we keep our focus on high levels, we assimilate them and thus push forth the evolutionary edge. We can also utilize the consciousness levels of other dimensions to bring about health, beauty and perfection here on the Earth. Mastery of natural and cosmic law may be in the primitive stage for us and yet be historic echoes in other dimensions which do not resist—once penetrated—divulging their secrets. Superhuman feats and abilities all stem from such a piercing. Einstein, Jung and Jefferson, to name just a few, all dreamed—all reached those expanded shores of consciousness. Why not us?

The sixth sense is inherent in each one of us. It lies dormant only because we have not understood its value to our everyday lives. If we seek to awaken it, it will become a valuable tool for knowledge of the self and others. Telepathy is simply a stillness that listens. If we meditate within the frequency of light, stillness ensues. Messages, like thoughts, appear on the surface to bring us answers and information from any direction to which we have opened our receptive mechanism.

Training our faculties to encompass even this one simple human attribute of telepathy could literally change our global future. Think what it would be like to know that your thoughts and intuitions are read by others. There could be no deception, therefore no victimization, no military or political coups, no covert anger or hatred! The certainty of exposure would discipline our thoughts to reflect only positive communication. Freed from the incessant "hum" of negative discourse, the mind would become the progenitor of beauty and love and oneness.

We don't need our old "King of the Mountain" strategy anymore and can release ourselves from reactive patterns to explore a new octave of merging heretofore nonexistent on this earth! We have experienced this merging, this universal oneness with cosmic consciousness in other dimensions. That illusive memory is still with us and provides the impetus to find it again. With past-life work we reexperience it energetically. The merging is actually with our divine source, yet once reunited with it, we can re-create it in all our interpersonal relationships. The ripple of this will so totally alter the nature of relationship that there will be no similarity with today's tentative, karmic interdependence. Once we know who we are and can see our karmic web of each relationship, we will be free to merge with each other as whole beings because we ourselves will be whole.

We can gain wonderful experiences, now, in the body. We can experience our light bodies, and occupy several places simultaneously. We could become so united with another that the feminine as well as the masculine principles are fully active. The gate to all of these powers is open to us. If we can see the interdependency and bonding between apparent external influences and inner perceptions, we can learn to master the realities of this earthly level. If we don't do it of our own accord, we'll be forced to by circumstances.

One cannot overlook that we are now caught in politically and ecologically precarious times. The threat of disintegration—of ourselves and our environment—due to our mindless pollution practices, or our foolish lust for atomic supremacy, has all the potential to manifest the prophecies foretold by every great seer and virtually all the world's religions that forecast "purification by fire." Chernobyl was a prophetic message to us all that technology is not a substitute for wisdom. In fact, Chernobyl was a warning about the illusion of separation. In this nuclear game, what happens to one happens to us all. How blatant the symbolic portent of our human future when mothers in Germany and other parts of Europe could not nurse their own babies because they had assimilated too much radioactive debris from wind patterns circulating around the globe.

Our present perceptions of self, and of our life and environment, fool us into believing that as individuals we are powerless; that we can't change external conditions; that we are victims of external and dissociated realities. Everything happening in the streets, in this country or anywhere on the Earth, is nothing other than a reflection of the themes and internal dilemmas that we must solve as individuals as well as collectively.

This test requires that we develop our inherent capabilities and manifest creative powers. We have come far enough along in our evolution to balance out disharmony and destruction if we can free the emotional body's addiction to them. We can even free ourselves from the shackles of deathly illnesses. If we align our consciousness with the entirety of our beingness, the source of all disease, which is spiritual, will become evident. We are capable of making quantum leaps in consciousness, so that collective illness is no longer necessary as a learning experience.

The German magazine *Esotera* quotes the wife (a doctor) of a former president of Germany, Frau Dr. Carstens, as saying, "To illustrate the effectiveness of a natural immunization of the body, I refer to 700 worldwide documented cases in which massive tumors and metastases were diagnosed hopeless by doctors, but disappeared without conventional healing." The relief to the body, to be sure, did not come from the material realm, but rather the shock to the body jolted it into communication with a higher will. The person who experiences the grace of such a connection to the true self emerges a new being: one who knows that he or she is part of a great manifesting force.

218 _____TIME IS AN ILLUSION

Meanwhile, there is adequate scientific knowledge at our disposal about the body processes, cell regeneration, spiritual, naturopathic and other healing methods, which can be effectively used to spare the body from acting out our emotional and spiritual imbalances. We should not delay any longer in creatively applying this data. If we ignore the indicators, we will be forced into a fight for survival.

It is like standing at the edge of a bed of glowing coals and thinking, "No, I can't walk across." And then suddenly being pushed forth by some internal force we didn't know we had.

We have already learned how to handle creative powers in Atlantis and other ancient civilizations. Materialization and other consciousness technologies, via energy manipulation to achieve material effects and forms, existed once on this earth to unimaginable extents. Unfortunately this collective memory of alchemy is still with us, and we feel the lust to control and manipulate matter. We were obsessed with personal power, which issued from those alchemic feats, and we still are. We perceived that energy called forth matter, but we ourselves became stuck in the formalities of those processes. We became embroiled in the ritualistic symbology of such power and eventually lost the potency necessary to manifest because of our seduction to the glamour of the "strutting" stance. Most rituals today suffer the same hollowness, a fleeting memory of past glory and nothing else. Nevertheless, rites and ritualistic objects were imbued with tremendous central force that entices us back into old layers of consciousness. They simply circle us around to the tenacious past. If we convince ourselves of our importance by the use of subconscious ritual, we will continually experience the powerful struggle to attain more and more personal power—with the same disastrous results!

This old energy of personal power is not appropriate now. It dealt with earlier steps in development, with demands for power, with victim-victimizer conflicts, with physical abuse and with many other entangling and confusing stages of passage. It isn't advantageous for us to become stuck in such forms or to reach for them again instead of using something symbolic that "represents" the real energy—we can access the higher mind and employ directly that essence which translates thought into matter. We need to reach deeper into the formless and let the light of consciousness show us the laws of divine manifestation.

It is direct, immediate, without external instruments. Only through that power lying within us all, which manifests itself in all dimensions, can we open and develop our consciousness in this way. The pure power of intent to know ourselves as complete beings will give us the answers to our daily lives because we can come into contact with the perfect blueprint of our choice and thus free ourselves from the burden of unconsciousness and its karmic result.

The future of this earth and all future worlds is already here in the unmanifest. Our challenge lies in all beings unfolding their inherent possibilities, for each and every one of us really knows about the inner divine power of light and sound; of creative energies. All we must do is consciously come into contact with it.

Naturally a polluted emotional body will want to restrain us. It will say, "That won't ever happen to me. One must have special talents that I don't possess." These and all the other recordings of separation, limitation and isolation seek to endlessly repeat. But there is great hope for the future if we use the consciousness available to us now in very practical terms.

Imagine what would happen if every morning you turned on the radio or TV and the announcer asked the entire audience to focus its mind's eye for thirty seconds on clear skies or reported a drought in some area and asked you to visualize it raining there. In Japan they have made significant headway in correcting vision difficulties by having whole factories stop and do eye exercises for ten minutes together. The power of focusing our intent is limitless. If children can call the rain, adults can dematerialize air pollution and even radiation. It's the same principle at work. The quality of all our thoughts interfaces with the environment in the same way as do any other pollutants. Our anger and fear poison everything around us. If we feel negative inside, we will witness aggression outside ourselves. Likewise, if we hone our thoughts to love and compassion, we will experience ourselves enveloped in divine communion with all things and all beings.

Not only does this herald a new octave of personal involvement in the world, but also a new avenue for our inherent sociability. Working together can bring such pleasure and accomplishment. There will be many crises to help us learn to focus on our oneness rather than our separateness. They will force us to handle this collectively. I sense a positive reversal in this area, not only in individual families and social groups but also between nations. When every single soul present on this plane brings his or her consciousness onto the global level, the "new age" will begin. We did not choose this game to pretend ourselves "wallflowers."

The source and goal of every person is perfection, oneness. These are not mental projections, but natural principles of the flow of life, which permeate everything. For this to become real for us we must step into the arena of the conscious search, the exploration of our multidimensional beingness. We must step in now!

If we witness others walking over hot coals, our frame of reference of "impossible" graduates to "possible." Our brain becomes inspired and responds to our own wondering if we can walk over hot coals with, "Yes, I, too, can do this." This book has a similar function: It should make clear that already so many other people have gone across the bridge into other dimensions, have jumped into the "unknown," can embrace themselves consciously and lovingly on the "other side," and then integrate those things into their daily lives to enrich themselves and others. Such things are possible. Time is an illusion! Experiencing multidimensionality lets us participate fully in our energy reality, awakening the power of choosing the "play" from a vast repertoire of possible combinations. We can dissolve and release the past and design the future by combining powerful energies borrowed from other, more expanded realms of consciousness. We can blend with any great being to increase our scope of truth.

You and I are souls of godly perfection. Let us discover it in ourselves—now!

Begin!

Information on seminars, lectures, and tapes by Chris Griscom can be obtained from:

The Light Institute of Galisteo
Rt. 3, Box 50
Galisteo, N.M. 87540
(505) 983-1975